Studies in Computational Intelligence

Volume 687

Series editor

Janusz Kacprzyk, Polish Academy of Sciences, Warsaw, Poland
e-mail: kacprzyk@ibspan.waw.pl

The series "Studies in Computational Intelligence" (SCI) publishes new developments and advances in the various areas of computational intelligence—quickly and with a high quality. The intent is to cover the theory, applications, and design methods of computational intelligence, as embedded in the fields of engineering, computer science, physics and life sciences, as well as the methodologies behind them. The series contains monographs, lecture notes and edited volumes in computational intelligence spanning the areas of neural networks, connectionist systems, genetic algorithms, evolutionary computation, artificial intelligence, cellular automata, self-organizing systems, soft computing, fuzzy systems, and hybrid intelligent systems. Of particular value to both the contributors and the readership are the short publication timeframe and the world-wide distribution, which enable both wide and rapid dissemination of research output.

More information about this series at http://www.springer.com/series/7092

J. K. Mandal · Paramartha Dutta
Somnath Mukhopadhyay
Editors

Advances in Intelligent Computing

 Springer

Editors
J. K. Mandal
Department of Computer Science
  and Engineering
University of Kalyani
Kalyani, West Bengal
India

Somnath Mukhopadhyay
Department of Computer
  Science and Engineering
Assam University
Silchar, Assam
India

Paramartha Dutta
Department of Computer
  and System Sciences
Visva-Bharati University
Santiniketan, Bolpur, West Bengal
India

ISSN 1860-949X   ISSN 1860-9503 (electronic)
Studies in Computational Intelligence
ISBN 978-981-13-4286-8   ISBN 978-981-10-8974-9 (eBook)
https://doi.org/10.1007/978-981-10-8974-9

Printed on acid-free paper

This Springer imprint is published by the registered company Springer Nature Singapore Pte Ltd. part of Springer Nature
The registered company address is: 152 Beach Road, #21-01/04 Gateway East, Singapore 189721, Singapore

Foreword

The neologism "Intelligent Computing" encompasses the theory, design, application and development of nature-inspired computational paradigms, emphasizing neural networks, connectionist systems, genetic algorithms, evolutionary programming, fuzzy systems, metaheuristic algorithms and hybrid intelligent systems in which these paradigms are contained. Intelligent computing is an emerging area of research and has hitherto yielded a rich crop of mathematical models that are being translated into a wealth of engineering applications in recent times.

Effective modelling of real-life multi-attribute decision-making, non-rigid registration of medical images, modelling of portfolio construction for Indian stock market, network-based intrusion detection, estimation of missing entries in the rating matrix of recommender systems, detection of primary user in cooperative spectrum sensing to enhance spectrum availability information for cognitive radio applications, recognition of handwritten words in Bengali, segregation of functionally inactive genes and noise, detection of moving objects in video sequences which have been outlined in this book exhibit that even imprecision and uncertainty could be effectively utilized in solving complex problems.

I am sure that the researchers will find in this book familiar as well as new landscapes in intelligent computing.

Bangalore, India
January 2018

Lalit M. Patnaik
INSA Senior Scientist and Adjunct Professor
Consciousness Studies Program
Indian Institute of Science
National Institute of Advanced Studies
http://www.lmpatnaik.in

Preface

The endeavour for realizing devices, systems having associated intelligence is one of the recent research trends. The genesis of development of so-called intelligence systems has its long past. By virtue of being naturally intelligent, human beings started exploring if some artificial analogue of intelligence is realizable. This effort was the advent of research in realization of "Intelligent Systems" with the crystallization of ingredients of artificial intelligence, courtesy research effort rendered by researchers in the domain, people started thinking about embedding these artificial intelligence components in passive computing system to convert it to somewhat an "intelligent Entity". This was the logical beginning of research effort towards machine intelligence. Different computational techniques used for the purpose of making a machine intelligent gradually became the focus of attention of the relevant research community. The formalization of computational techniques used for the purpose of achieving intelligent systems is what computational intelligence all about.

It is a small but sincere effort on the part of the present editors to offer a volume where different aspects of computational intelligence have been reported. The objective of this publication is to enlighten the researcher, scholars, students and engineers about the state-of-the-art scenario regarding advances in intelligent computing techniques and associated intelligence paradigm, the latest tools and techniques which are applicable to almost all leading fields of current research. The theme is illustrated in various chapters to encourage researchers to adopt it in multidisciplinary research and engineering design. We hope that promising ideas and outstanding research results of this issue will instil further development of research and will enhance technologies in terms of methodologies and applications of computational intelligence.

This edited book volume entitled *Advances in Intelligent Computing* is a collection of eight chapters which are post-conference publications as extended version papers from the first international conference on Computational Intelligence, Communications, and Business Analytics (CICBA 2017), held at Kolkata, during

24–25 March 2017 under the publication house of Springer Nature Singapore in CCIS series. The chapters were initially peer-reviewed by the Editorial Review Board members, and reviewers who themselves span over many countries. A brief description of each of the chapters is as follows:

Chapter "Linear Programming-Based TOPSIS Method for Solving MADM Problems with Three Parameter IVIFNs" deals with developing a TOPSIS approach using fractional programming techniques for effective modelling of real-life multi-attribute decision-making (MADM) problems in interval-valued intuitionistic fuzzy (IVIF) settings by considering hesitancy degree as a dimension together with membership and non-membership degrees. In three-parameter characterization of intuitionistic fuzzy (IF) sets, a weighted absolute distance between two IF sets with respect to IF weights is defined and employed in TOPSIS to formulate intervals of relative closeness coefficients (RCCs). The lower and upper bounds of the intervals of RCCs are given by a pair of nonlinear fractional programming models which are further transformed into two auxiliary linear programming models using mathematical methods and fractional programming technique. A simpler technique is also proposed for estimating the optimal degrees as performance values of alternatives from the possibility degree matrix generated by pairwise comparisons of RCC intervals. The validity and effectiveness of the proposed approach are demonstrated in the chapter through some numerical examples.

In Chapter "A Comparative Study of Bio-inspired Algorithms for Medical Image Registration", the challenge of determining optimal transformation parameters for image registration has been treated traditionally as a multidimensional optimization problem. Non-rigid registration of medical images has been approached in this chapter using the particle swarm optimization algorithm, dragonfly algorithm and the artificial bee colony algorithm. Brief introductions to these algorithms have been presented in the chapter. Results of medical image registration using the proposed algorithms have been analysed in the chapter. The simulation shows that the dragonfly algorithm results in higher quality image registration, but takes longer to converge. The trade-off issue between the quality of registration and the computing time has been brought forward.

In Chapter "Different Length Genetic Algorithm-Based Clustering of Indian Stocks for Portfolio Optimization", authors proposed a model for portfolio construction using Different Length Genetic Algorithm (GA)-based clustering of Indian stocks. Stocks of different companies, chosen from different industries, are classified on their returns per unit of risk using an unsupervised method of Different Length Genetic Algorithm. After that the obtained centroids are again classified by the proposed algorithm. Vertical clustering (clustering of stocks by returns per unit of risk for each day) followed by horizontal clustering (clustering of the centroids over time) eventually produces a limited number of stocks. The Markowitz model is then applied to determine the weights of the stocks in the portfolio. The results are also compared with some well-known existing algorithms.

In Chapter "An Evolutionary Matrix Factorization Approach for Missing Value Prediction", authors addressed the issue of performance degradation of fuzzy c-means (FCM) clustering. They stated that FCM suffers from low signal-to-noise ratio (SNR) due to the non-spherical nature of the dataset developed from the energy values of the sensed signal. To address this problem, authors explored the scope of possibilistic fuzzy c-means (PFCM) algorithm on energy detection-based spectrum sensing (SS). PFCM combines the fuzzy membership function and the possibilistic information in the clustering process to partition the inseparable energy data into the respective clusters. Differential evolution (DE) algorithm is applied along with PFCM to maximize the probability of PU detection (PD) under the constraint of a target false alarm probability (PFA).

In Chapter "Differential Evolution in PFCM Clustering for Energy Efficient Cooperative Spectrum Sensing", authors proposed a Memetic Algorithm (MA)-based wrapper-filter feature selection method which is applied for the recognition of handwritten words' images in segmentation-free approach. They considered two state-of-the-art feature vectors describing texture and shape of the word images, respectively, for feature dimension reduction. Authors have shown experimental results on handwritten Bangla word samples comprising 50 popular city names of West Bengal, a state of India.

In Chapter "Feature Selection for Handwritten Word Recognition Using Memetic Algorithm", authors proposed a clustering method that clusters co-expressed genes and segregates functionally inactive genes and noise. The proposed method formed a cluster if the difference in expression levels of genes with a specified gene is less than a threshold t in each experiment condition; otherwise, the specified gene is marked as functionally inactive or noise. The proposed method is applied on 10 yeast gene expression data and the result shows that it performs well over existing one.

In Chapter "A Column-Wise Distance-Based Approach for Clustering of Gene Expression Data with Detection of Functionally Inactive Genes and Noise", authors have proposed a technique to detect moving objects in the video under dynamic as well as static background condition. The method consists block-based background modelling, current frame updating, block processing of updated current frame and elimination of background using bin histogram approach. Then the enhanced foreground objects are obtained in the post-processing stage using morphological operations. They have shown that the approach effectively minimizes the effect of dynamic background to extract the foreground information. They have applied the technique on Change Detection CDW-2012 dataset and compared the results with the other state-of-the-art methods.

In Chapter "Detection of Moving Objects in Video Using Block-Based Approach", authors proposed an efficient technique for detecting moving objects in the video under dynamic as well as static background condition. The method consists block-based background modelling, current frame updating, block processing of updated current frame and elimination of background using bin histogram approach. After that an enhanced foreground objects are obtained in the post-processing stage using morphological operations. The proposed approach effectively minimizes the effect of dynamic background to extract the foreground information. Authors have applied the proposed technique on Change Detection CDW-2012 dataset and compared the results with the other state-of-the-art methods. The experimental results prove the efficiency of the proposed approach compared to the other state-of-the-art methods in terms of different evaluation metrics.

In conclusion, this volume is composed with maximum nourishing aiming to shield all main aspects of the area of intelligent computing with proposing compact background information as well as end-point applications from the bests in the area reinforced by younger investigators in the field. It is designed to be a one-stop-shop for interested readers, but by no means aims to completely interchange all other sources in this vigorously developing area of research.

Enjoy reading it.

Kalyani, India   J. K. Mandal
Santiniketan, India   Paramartha Dutta
Silchar, India   Somnath Mukhopadhyay

Contents

About the Editors

J. K. Mandal has degrees M.Sc.(Ph.), JU; M.Tech.(CS), CU; Ph.D.(Eng.), JU; and is a Professor in the Department of CSE. He is former Dean of FETM and was KU for two consecutive terms. He has teaching and research experience spanning 29 years and completed four AICTE and one state government project. He is a Life Member of CSI, CRSI; a Member of ACM; and a Fellow of IETE. He was also Honorary Vice-Chairman and Chairman of CSI. He has delivered over 100 lectures and organized more than 25 national and international conferences. He has produced 20 Ph.D.'s; three scholars have submitted; and eight are pursuing research under his guidance. He has supervised three M.Phil. and more than 50 M.Tech. and 100 M.C.A. dissertations. He is an editorial board member and corresponding editor of the Proceedings of ScienceDirect, IEEE and other conferences as well as Guest Editor of the MST Journal. He has published more than 400 research articles and six books.

Paramartha Dutta was born in 1966 and has B.Stat., M.Stat. and M.Tech. degrees in Computer Science, all from the Indian Statistical Institute, and a Ph.D. (Engineering) from the Bengal Engineering and Science University, Shibpur. He served as a researcher in various projects funded by Government of India. He is now a Professor at the Department of Computer and System Sciences, Visva-Bharati University, West Bengal, India. Prior to this, he worked at the Kalyani Government Engineering College and College of Engineering in West Bengal. He was a visiting/guest scholar at several universities/institutes such as West Bengal University of Technology, Kalyani University and Tripura University. He has co-authored eight books and has six edited books to his credit. He has published over 200 papers in various international and national journals and conference proceedings, as well as several chapters in edited volumes. He has served as the editor of special volumes of several respected international journals.

Somnath Mukhopadhyay is currently an Assistant Professor at Department of Computer Science and Engineering, Assam University, Silchar, India. He completed his M.Tech. and Ph.D. degrees in Computer Science and Engineering at the

University of Kalyani, India, in 2011 and 2015, respectively. He has co-authored one book and has five edited books to his credit. He has published over 20 papers in various international journals and conference proceedings, as well as three chapters in edited volumes. His research interests include digital image processing, computational intelligence and pattern recognition. He is a member of IEEE and IEEE Young Professionals, Kolkata Section; life member of the Computer Society of India; and currently the regional student coordinator (RSC) of Region II, Computer Society of India.

Linear Programming-Based TOPSIS Method for Solving MADM Problems with Three Parameter IVIFNs

Samir Kumar and Animesh Biswas

Abstract The aim of this paper is to develop a TOPSIS approach using fractional programming techniques for effective modelling of real-life multiattribute decision-making (MADM) problems in interval-valued intuitionistic fuzzy (IVIF) settings by considering hesitancy degree as a dimension together with membership and non-membership degrees. In three-parameter characterizations of intuitionistic fuzzy (IF) sets, a weighted absolute distance between two IF sets with respect to IF weights is defined and employed in TOPSIS to formulate intervals of relative closeness coefficients (RCCs). The lower and upper bounds of the intervals of RCCs are given by a pair of nonlinear fractional programming models which are further transformed into two auxiliary linear programming models using mathematical methods and fractional programming technique. A simpler technique is also proposed for estimating the optimal degrees as performance values of alternatives from the possibility degree matrix generated by pairwise comparisons of RCC intervals. The validity and effectiveness of the proposed approach are demonstrated through two numerical examples.

Keywords Intuitionistic fuzzy sets · Interval-valued intuitionistic fuzzy numbers
TOPSIS · Mathematical programming · Possibility degree matrix

S. Kumar
Department of Mathematics, Acharya Jagadish Chandra Bose College,
Kolkata 700020, West Bengal, India
e-mail: kumarsamir2007@gmail.com

A. Biswas (✉)
Department of Mathematics, University of Kalyani, Kalyani,
Nadia 741235, West Bengal, India
e-mail: abiswaskln@rediffmail.com

© Springer Nature Singapore Pte Ltd. 2019
J. K. Mandal et al. (eds.), *Advances in Intelligent Computing*,
Studies in Computational Intelligence 687,
https://doi.org/10.1007/978-981-10-8974-9_1

1

1 Introduction

Multiattribute decision-making (MADM) refers to making selections among some courses of action in the presence of multiple conflicting attributes, and it needed to employ the available scientific tools and techniques to accurately model real-life unstructured decision problems for better and effective solutions. The technique for order preference by similarity to ideal solution (TOPSIS) technique as a very significant MADM method was proposed by Hwang and Yoon [21]. The positive ideal solution (PIS) and the negative ideal solution (NIS) are constructed as combinations of the best and the worst performance values of each alternative relative to each criterion as represented by the decision matrix. Hence, the PIS and the NIS are the best and the worst possible alternatives, respectively. The fundamental principle of TOPSIS is that the most desirable alternative should be nearest to the PIS and farthest from the NIS. The distance is measured by using appropriate metric in a given decision system. Chen [11] extended TOPSIS in fuzzy environment by employing the concepts of fuzzy set theory (FST) [30]. FST based on membership function established its effectiveness [6, 8, 9] in the field of modelling [5, 7] uncertainty and vagueness, whereas intuitionistic fuzzy sets (IFSs) are characterized by both membership and non-membership functions. Atanassov [1] introduced IFSs as a generalization of fuzzy sets [30] for capturing imprecision through processing the data on the basis of membership and non-membership functions. Later, Gau and Buehrer [17] proposed the concepts of vague sets as another generalization of fuzzy sets. The theory of vague sets has been used a number of times in dealing with multicriteria fuzzy decision-making problems [12, 20]. However, Bustince and Burillo [10] established that vague sets are identical with IFSs. Thus, the theory of IFSs remained superior tools for modelling indecisiveness and incompleteness in numerous decision contexts in comparison with fuzzy sets [3, 4, 23, 24, 28, 31, 32]. Motivated by the notion of interval-valued fuzzy sets, Atanassov and Gargov [2] introduced the concepts of interval-valued IFSs (IVIFSs) by generalizing IFSs bearing greater power for dealing with uncertainties in the real-world problems. Deschrijver and Kerre [15] made a very systematic study about the mathematical relationships between IFSs and various models of imprecision. Deschrijver and Kerre [14] studied the relationships among IFSs, L-fuzzy sets, interval-valued fuzzy sets and IVIFSs. Based on the concepts of inclusion comparison possibilities, Chen [13] extended the TOPSIS in interval-valued intuitionistic fuzzy (IVIF) settings by developing an index for an inclusion-based closeness coefficient. Zeng and Xiao [31] proposed a hybrid intuitionistic fuzzy (IF) ordered weighted averaging weighted averaging (OWAWA) distance formula and presented a TOPSIS method for IF multicriteria decision-making problems. Zhao [32] developed an optimization model for determination of criteria weights and advanced a TOPSIS method in IVIF ambience using weighted Hamming distance between each alternative and PIS and NIS for solving MADM problems. The tools and techniques of programming methodology have been combined with the concepts of IFSs [23, 24] for solving MADM problems. Li [24] proposed weighted absolute distance between two IFSs with respect to IF weights by considering two parameter characterizations of IFSs and applied mathematical programming techniques to solve MADM problems. Biswas and Kumar [4] developed mathematical-programming-based TOPSIS techniques using three-parameter characterizations of IFSs and normalized hamming distance in IVIF settings. Szmidt and Kacprzyk [26] demonstrated that the third parameter intuitionistic index cannot be omitted from the definition of distance between two IFSs and defined the four basic distances, the Hamming distance, the normalized Hamming distance, the Euclidean distance and the normalized Euclidean distance, between the IFSs with the three three-parameter characterizations (membership, non-membership and hesitancy degrees) of IFSs. Xu and Yager [28] further proposed distance between two IF numbers (IFNs) in terms of three parameter characterizations.

In this paper, due to assertions regarding the presence of hesitancy degree in the formulations of distances [26, 28] between IFSs in three parameter characterizations, the work of Li [24] is generalized by proposing an absolute distance between two IFSs with respect to IF weights by considering the three parameter characterization of IFSs and used in TOPSIS to formulate intervals of relative closeness coefficients (RCCs) for yielding better results. The lower and upper bounds of RCC intervals are obtained by developing a pair of nonlinear fractional programming (NLFP) models. The NLFP models are transformed into two auxiliary linear programming models using mathematical methods and fractional programming techniques. A simpler formula is also advanced for estimating the performance values of alternatives from the possibility degree matrix generated by pairwise comparisons of RCC intervals.

The rest of the paper is organized as follows. Section 2 reviews the concepts of interval-valued IFNs (IVIFNs) and possibility degree for ranking interval numbers. Section 3 defines an MADM problem and a weighted absolute distance between two IFSs with respect to the IF weights with three-parameter characterizations of IFSs. In Sect. 4, the bounds of intervals of RCCs under TOPSIS approach are obtained. Then, the possibility degree matrix is obtained by pairwise comparison of RCC intervals and a revised formula for optimal degrees is used to get the performance values of alternatives yielding their ranking. The developed approach is illustrated through two numerical examples in Sect. 5 while concluding remarks are made in Sect. 6.

2 Preliminaries

Atanassov and Gargov [2] generalized IFSs to IVIFSs by allowing the membership and non-membership functions to assume subintervals of the unit interval [0, 1] as their values at the place of exact numbers and hence consistent with the human judgment in uncertain environment.

The concepts of IVIFSs and the possibility degree measures for comparing two interval-valued fuzzy numbers are recapitulated in this section.

Definition 2.1 [2] Assume that $X(\neq\varnothing)$ be a universe of discourse and $D[0,1]$ denotes the set of all closed subintervals of $[0,1]$. An IVIFS, A, on X is defined as a set $A=\{\langle x,\mu_A(x),\nu_A(x)\rangle:x\in X\}$, where $\mu_A:X\to D[0,1]$ and $\nu_A:X\to D[0,1]$, respectively, represent membership and non-membership functions satisfying the condition $0\le sup\mu_A(x)+sup\nu_A(x)\le 1$, for each $x\in X$.

The interval degrees of membership and non-membership of an element $x(\in X)$ to the set A are denoted, respectively, by $\mu_A(x)$ and $\nu_A(x)$. For each $x\in X$, assume that the infimum and supremum of $\mu_A(x)$ and $\nu_A(x)$ are represented, respectively, by $\mu_A^l(x),\mu_A^u(x)$ and $\nu_A^l(x),\nu_A^u(x)$. Hence, the IVIFS A can be expressed as $A=\{\langle x,[\mu_A^l(x),\mu_A^u(x)],[\nu_A^l(x),\nu_A^u(x)]\rangle:x\in X\}$ satisfying the condition $0\le\mu_A^u(x)+\nu_A^u(x)\le 1,\mu_A^l(x)\ge 0,\nu_A^l(x)\ge 0$.

For IVIFS A in X, an associated function $\pi_A:X\to D[0,1]$ is defined by

$$\pi_A(x)=1-\mu_A(x)-\nu_A(x)$$
$$=\left[\pi_A^l(x),\pi_A^u(x)\right]=\left[1-\mu_A^u(x)-\nu_A^u(x),1-\mu_A^l(x)-\nu_A^l(x)\right],\forall x\in X$$

The functional value $\pi_A(x)$ represents the interval hesitancy degree of an element x in A.

For each $x\in X$, the ordered triple $\langle\mu_A(x),\nu_A(x),\pi_A(x)\rangle$ was called an interval-valued intuitionistic fuzzy number (IVIFN) by Xu and Yager [28]. For brevity, an IVIFN is represented by $A=\langle\mu_A,\nu_A,\pi_A\rangle$.

As there is no scope for hesitation in fuzzy sets [30], the decision makers are constrained to make complimentary choices of membership and non-membership degrees while the three-parameter characterizations [26] of the IFS A assigns a membership degree $\mu_A(x)$, a non-membership degree $\nu_A(x)$ and a hesitancy degree $\pi_A(x)$ [4, 26, 28] to each element $x\in X$ giving greater liberty to decision makers for better and effective opinions in a given decision context. Thus, an element $\alpha\in A(\text{IFS})$ can be expressed as $\alpha=\langle\mu_\alpha,\nu_\alpha,\pi_\alpha\rangle$ having three coordinates. An IFS can be well understood in three-dimensional figure (Fig. 1) by drawing a unit cube with the $\triangle ABC$ representing IFS while $\triangle OAB$ is an orthogonal projection of $\triangle ABC$ (real situation). The points $A\langle 1,0,0\rangle,B\langle 0,1,0\rangle,$ and $C\langle 0,0,1\rangle$ represent intuitionistic fuzzy numbers which correspond, respectively, to absolute acceptance, total rejection and full hesitation regarding the choice of a decision maker. The line segment AB represents a fuzzy set [26].

It is obvious that $A^+=\langle[1,1],[0,0],[0,0]\rangle$ and $A^-=\langle[0,0],[1,1],[0,0]\rangle$ represent the greatest and smallest IVIFNs, respectively.

The imprecision and uncertainty in nature are better captured through interval numbers than exact real numbers. Let $a_i=\left[a_i^l,a_i^u\right](i=1,2)$ be two interval numbers. Obviously, it is difficult to define an order $a_1\ge a_2$ between the interval numbers a_1 and $a_2,$ which has fuzzy characteristics. The concept of possibility degree [16, 22] for comparing interval numbers is defined as follows:

Definition 2.2 [22] Let $a_i=\left[a_i^l,a_i^u\right](i=1,2)$ be two interval numbers. Then, the possibility degree of $a_1\ge a_2$ is defined as the degree of compatibility of $a_1\ge a_2$ with

Fig. 1 Three-dimensional representation of an intuitionistic fuzzy set

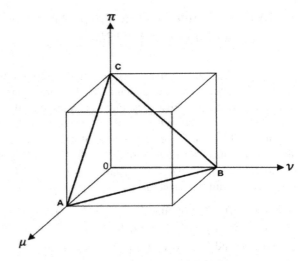

the truthfulness of the proposition "a_1 is not smaller than a_2" and is expressed by the membership degree $\mu_T(a_1 \geq a_2)$.

For convenience, the possibility degree of $a_1 \geq a_2$ is denoted $p(a_1 \geq a_2)$, i.e. $\mu_T(a_1 \geq a_2) = p(a_1 \geq a_2)$.

There are many formulations for estimating the possibility degree $p(a_1 \geq a_2)$ of $a_1 \geq a_2$. However, for ease of computation, the following definition is considered here.

Definition 2.3 [16] Let $a_i = \left[a_i^l, a_i^u\right] (i = 1, 2)$ be two interval numbers with interval lengths $L(a_i) = a_i^u - a_i^l (i = 1, 2)$. Then, the possibility degree $p(a_1 \geq a_2)$ of $a_1 \geq a_2$ is defined by

$$p(a_1 \geq a_2) = min\left\{ max\left\{ \frac{a_1^u - a_2^l}{L(a_1) + L(a_2)}, 0 \right\}, \ 1 \right\}$$

Clearly, the degree of possibility $p(a_1 \geq a_2)$ of $a_1 \geq a_2$ satisfies the following important properties:

(i) $0 \leq p(a_1 \geq a_2) \leq 1$;

(ii) $p(a_1 \geq a_2) + p(a_2 \geq a_1) = 1$;

(iii) $p(a_1 \geq a_1) = 0.5$;

(iv) $p(a_1 \geq a_2) = 1$ if and only if $a_1^l \geq a_2^u$;

(v) $p(a_1 \geq a_2) = 0$ if and only if $a_1^u \leq a_2^l$;

(vi) $p(a_1 \geq a_2) \geq 0.5$ if and only if $a_1^l + a_1^u \geq a_2^l + a_2^u$;

(vii) Assume that $a_i = \left[a_i^l, a_i^u\right] (i = 1, 2, 3)$ are three interval numbers. If $p(a_1 \geq a_2) \geq 0.5$ and $p(a_2 \geq a_3) \geq 0.5$ then $p(a_1 \geq a_3) \geq 0.5$

3 MADM Problems with Three Parameter Representations of IVIFNs

Szmidt and Kacprzyk [26] established that any extension of fuzzy distance measure in the IF settings should contain all the three parameters, viz. membership, non-membership and hesitancy degrees. Subsequently, Xu and Yager [28] proposed the distance between two IFNs by considering three parameter characterizations of IFNs for better modelling of uncertainty. In this article, a weighted absolute distance between two IFSs with respect to IF weights is defined by employing the three-parameter characterizations of IFSs as an extension of the weighted absolute distance described by Li [24].

Definition 3.1. Assume that $X = \{x_1, x_2, \ldots, x_n\}$ be a finite universe of discourse. Let $A = \{\langle \mu_A(x_j), \nu_A(x_j), \pi_A(x_j)\rangle : x_j \in X\}$, $B = \{\langle \mu_B(x_j), \nu_B(x_j), \pi_B(x_j)\rangle : x_j \in X\}$ be two IFSs and $M = \{\langle \mu_M(x_j), \nu_M(x_j), \pi_M(x_j)\rangle : x_j \in X\}$ be an IFS representing the weights of importance of all the elements $x_j \in X$.

Then, a weighted absolute distance between IFSs A and B with respect to the IFS M is defined as

$$d(A,B) = \sum_{j=1}^{n} \left[\mu_M(x_j)\left|\mu_A(x_j) - \mu_B(x_j)\right| + \nu_M(x_j)\left|\nu_A(x_j) - \nu_B(x_j)\right| + \pi_M(x_j)\left|\pi_A(x_j) - \pi_B(x_j)\right| \right]$$

$$(1)$$

Now, this concept has been extended for IVIFSs to find best alternative in MADM context.

3.1 Linear Programming-Based TOPSIS Approach in Three Parameter IVIF Settings

Assume that a decision maker evaluates m alternatives $A_i(i = 1, 2, \ldots, m)$ based on n criteria $\Gamma_j(j = 1, 2, \ldots, n)$ in an MADM problem and expresses judgement values in the form of IVIFNs generating the three parameter IVIF decision matrix as
$D = (\alpha_{ij})_{m \times n} = (\langle[\mu_{ij}^l, \mu_{ij}^u], [\nu_{ij}^l, \nu_{ij}^u], [\pi_{ij}^l, \pi_{ij}^u]\rangle)_{m \times n} =$

$$
\begin{array}{c}
A_1 \\ A_2 \\ \vdots \\ A_m
\end{array}
\begin{pmatrix}
\langle[\mu_{11}^l, \mu_{11}^u], [\nu_{11}^l, \nu_{11}^u], [\pi_{11}^l, \pi_{11}^u]\rangle & \langle[\mu_{12}^l, \mu_{12}^u], [\nu_{12}^l, \nu_{12}^u], [\pi_{12}^l, \pi_{12}^u]\rangle & \cdots & \langle[\mu_{1n}^l, \mu_{1n}^u], [\nu_{1n}^l, \nu_{1n}^u], [\pi_{1n}^l, \pi_{1n}^u]\rangle \\
\langle[\mu_{21}^l, \mu_{21}^u], [\nu_{21}^l, \nu_{21}^u], [\pi_{21}^l, \pi_{21}^u]\rangle & \langle[\mu_{22}^l, \mu_{22}^u], [\nu_{22}^l, \nu_{22}^u], [\pi_{22}^l, \pi_{22}^u]\rangle & \cdots & \langle[\mu_{2n}^l, \mu_{2n}^u], [\nu_{2n}^l, \nu_{2n}^u], [\pi_{2n}^l, \pi_{2n}^u]\rangle \\
\vdots & \vdots & & \vdots \\
\langle[\mu_{m1}^l, \mu_{m1}^u], [\nu_{m1}^l, \nu_{m1}^u], [\pi_{m1}^l, \pi_{m1}^u]\rangle & \langle[\mu_{m2}^l, \mu_{m2}^u], [\nu_{m2}^l, \nu_{m2}^u], [\pi_{m2}^l, \pi_{m2}^u]\rangle & \cdots & \langle[\mu_{mn}^l, \mu_{mn}^u], [\nu_{mn}^l, \nu_{mn}^u], [\pi_{mn}^l, \pi_{mn}^u]\rangle
\end{pmatrix}
$$

with column headers Γ_1, Γ_2, $\ldots$, Γ_n.

Further, assume that the decision maker also evaluates criteria $\Gamma_j(j = 1, 2, \ldots, n)$ generating three parameter IVIF criteria weight vector

$w = (w_j)_{1 \times n}$, where $w_j = \langle \left[\omega_j^l, \omega_j^u \right], \left[\rho_j^l, \rho_j^u \right], \left[\sigma_j^l, \sigma_j^u \right] \rangle$ is the IVIFN weight for the jth criterion Γ_j.

For each criterion $\Gamma_j (j = 1, 2, \ldots, n)$, let α_j^+ and α_j^-, be the respective greatest and the smallest IVIFNs in the decision matrix D corresponding to all the alternatives $A_i (i = 1, 2, \ldots, m)$ which are identified as the jth component of the interval-valued intuitionistic fuzzy PIS (IVIF-PIS) and interval-valued intuitionistic fuzzy NIS (IVIF-NIS), respectively, with respect to the criterion Γ_j. Thus, the IVIF-PIS α^+ and IVIF-NISs α^- are the best and worst values of each alternative $A_i (i = 1, 2, \ldots, m)$ relative to all criteria $\Gamma_j (j = 1, 2, \ldots, n)$ which are expressed as follows:

$$\alpha^+ = (\alpha_1^+, \alpha_2^+, \ldots, \alpha_n^+), \text{where } \alpha_j^+ = \langle [1, 1], [0, 0], [0, 0] \rangle, \text{for } j = 1, 2, \ldots, n$$

$$\alpha^- = (\alpha_1^-, \alpha_2^-, \ldots, \alpha_n^-), \text{where } \alpha_j^- = \langle [0, 0], [1, 1], [0, 0] \rangle, \text{for } j = 1, 2, \ldots, n$$

Assume now that $\mu_{ij} \in \left[\mu_{ij}^l, \mu_{ij}^u \right], \nu_{ij} \in \left[\nu_{ij}^l, \nu_{ij}^u \right], \pi_{ij} \in \left[\pi_{ij}^l, \pi_{ij}^u \right]$ and $\omega_j \in \left[\omega_j^l, \omega_j^u \right], \rho_j \in \left[\rho_j^l, \rho_j^u \right], \sigma_j \in \left[\sigma_j^l, \sigma_j^u \right]$. Thus, from the decision matrix D in the given MADM problem, it is found that for each alternative $A_i (i = 1, 2, \ldots, m)$, there exists an n-component IF vector $\beta_i = (\beta_{ij})_{1 \times n} (i = 1, 2, \ldots, m)$, where $\beta_{ij} = \langle \mu_{ij}, \nu_{ij}, \pi_{ij} \rangle$.

Now, the newly defined distance formula as given in Eq. (1) is used in TOPSIS approach to find RCC by evaluating the weighted absolute distance between the IF vector β_i, corresponding to ith alternative $A_i (i = 1, 2, \ldots, m)$, and the IVIF-PIS α^+ with respect to all criteria $\Gamma_j (j = 1, 2, \ldots, n)$ as follows:

$$d(\beta_i, \alpha^+) = \sum_{j=1}^{n} \left[\omega_j (1 - \mu_{ij}) + \rho_j \nu_{ij} + \sigma_j \pi_{ij} \right] \tag{2}$$

where $\pi_{ij} = 1 - \mu_{ij} - \nu_{ij}$ and $\sigma_j = 1 - \omega_j - \rho_j$

In the like manner, using Eq. (1), the weighted absolute distance between the IF vector β_i and the IVIF-NIS α^- with respect to all criteria $\Gamma_j (j = 1, 2, \ldots, n)$, is expressed as

$$d(\beta_i, \alpha^-) = \sum_{j=1}^{n} \left[\omega_j \mu_{ij} + \rho_j (1 - \nu_{ij})) + \sigma_j \pi_{ij} \right] \tag{3}$$

where $\pi_{ij} = 1 - \mu_{ij} - \nu_{ij}$ and $\sigma_j = 1 - \omega_j - \rho_j$

Using Eqs. (2) and (3), the RCCs of alternatives $A_i (i = 1, 2, \ldots, m)$ with respect to the IVIF-PIS α^+ are defined as follows:

$$C_i(\mu_{ij}, \nu_{ij}, \pi_{ij}) = \frac{d(\beta_i, \alpha^-)}{d(\beta_i, \alpha^+) + d(\beta_i, \alpha^-)}, \quad \text{for } i = 1, 2, \ldots, m$$

$$\text{i.e., } C_i = \frac{\sum_{j=1}^{n} [\omega_j \mu_{ij} + \rho_j (1 - \nu_{ij}) + \sigma_j \pi_{ij}]}{\sum_{j=1}^{n} [\omega_j (1 - \mu_{ij}) + \rho_j \nu_{ij} + \sigma_j \pi_{ij}] + \sum_{j=1}^{n} [\omega_j \mu_{ij} + \rho_j (1 - \nu_{ij}) + \sigma_j \pi_{ij}]}$$

$$= \frac{\sum_{j=1}^{n} [\omega_j \mu_{ij} + \rho_j (1 - \nu_{ij}) + \sigma_j \pi_{ij}]}{\sum_{j=1}^{n} [\omega_j + \rho_j + 2\sigma_j \pi_{ij}]}$$

$$\Rightarrow C_i = \frac{\sum_{j=1}^{n} [\omega_j \mu_{ij} + \rho_j (1 - \nu_{ij}) + \sigma_j (1 - \mu_{ij} - \nu_{ij})]}{\sum_{j=1}^{n} [\omega_j + \rho_j + 2\sigma_j (1 - \mu_{ij} - \nu_{ij})]} \qquad [\text{Since } \pi_{ij} = 1 - \mu_{ij} - \nu_{ij}]$$

$$\text{where } \sigma_j = 1 - \omega_j - \rho_j$$

$$= \frac{R_i}{S_i} \quad (\text{say}), \quad \text{for } i = 1, 2, \ldots, m$$

$$\Rightarrow C_i = \frac{R_i}{S_i}, \quad \text{for } i = 1, 2, \ldots, m$$

$$(4)$$

Taking logarithm of both sides of Eq. (4),

$$\log C_i = \log R_i - \log S_i \qquad (5)$$

Partial derivatives of Eq. (5) with respect to $\mu_{ik}(k = 1, 2, \ldots, n)$ are estimated as

$$\frac{\partial \log C_i}{\partial \mu_{ik}} = \frac{\omega_k - \sigma_k}{R_i} + \frac{2\sigma_k}{S_i} \geq \frac{\omega_k - \sigma_k}{S_i} + \frac{2\sigma_k}{S_i} \quad [\because R_i \leq S_i \text{ and assuming } \omega_k \geq \sigma_k]$$

$$= \frac{\omega_k + \sigma_k}{S_i} \geq 0$$

Since $\omega_k \in [\omega_k^l, \omega_k^u] \subseteq [0, 1], \rho_k \in [\rho_k^l, \rho_k^u] \subseteq [0, 1]$ and $\sigma_k \in [\sigma_k^l, \sigma_k^u] \subseteq [0, 1]$

Therefore, the RCC function C_i is a monotonically increasing function with respect to μ_{ik} as $\log C_i$ is a monotonically increasing function with respect to μ_{ik}.

Now, partial derivatives of Eq. (5) with respect to $\nu_{ik}(k = 1, 2, \ldots, n)$ are estimated as

$$\frac{\partial \log C_i}{\partial \nu_{ik}} = \frac{-\rho_k - \sigma_k}{R_i} + \frac{2\sigma_k}{S_i} \leq \frac{-\rho_k - \sigma_k}{R_i} + \frac{2\sigma_k}{R_i} = -\frac{(\rho_k - \sigma_k)}{R_i} \leq 0$$

$$[\because R_i \leq S_i \text{ and assuming } \rho_k \geq \sigma_k]$$

Therefore, the RCC function C_i is a monotonically decreasing function with respect to ν_{ik} as $\log C_i$ is a monotonically decreasing function with respect to ν_{ik}.

Remark 3.1 In the previous theoretical development, the consideration of inequalities $\omega_k \geq \sigma_k$ and $\rho_k \geq \sigma_k (k = 1, 2, \ldots, n)$ can be explained from the fact that the unknown three parameter intuitionistic fuzzy number criterion weight $\langle \omega_k, \rho_k, \sigma_k \rangle$ with $\omega_k \in \left[\omega_k^l, \omega_k^u\right], \rho_k \in \left[\rho_k^l, \rho_k^u\right], \sigma_k \in \left[\sigma_k^l, \sigma_k^u\right]$ in IVIF MADM; the solution remains meaningful by allowing hesitancy degree (σ_k) smaller than both membership (ω_k) and non-membership degree (σ_k). The lesser hesitancy degree in an IVIF decision-making problem, the better is the solution.

As the values of the RCC function C_i depend on the values $\mu_{ij} \in \left[\mu_{ij}^l, \mu_{ij}^u\right]$, $\nu_{ij} \in \left[\nu_{ij}^l, \nu_{ij}^u\right]$, $\omega_j \in \left[\omega_j^l, \omega_j^u\right], \rho_j \in \left[\rho_j^l, \rho_j^u\right], \sigma_j \in \left[\sigma_j^l, \sigma_j^u\right], C_i$ should be an interval.

The bounds of the interval of RCC C_i are estimated using the following NLFP models:

Find $\left(\mu_{ij}, \nu_{ij}, \omega_j, \rho_j, \sigma_j\right)$

so as to

$$Max\ C_i = \frac{\sum_{j=1}^{n}\left[\omega_j \mu_{ij} + \rho_j(1 - \nu_{ij}) + \sigma_j\left(1 - \mu_{ij} - \nu_{ij}\right)\right]}{\sum_{j=1}^{n}\left[\omega_j + \rho_j + 2\sigma_j\left(1 - \mu_{ij} - \nu_{ij}\right)\right]}$$

$$\text{subject to} \begin{cases} \mu_{ij}^l \leq \mu_{ij} \leq \mu_{ij}^u \ (i = 1, 2, \ldots, m; j = 1, 2, \ldots, n), \\ \nu_{ij}^l \leq \nu_{ij} \leq \nu_{ij}^u \ (i = 1, 2, \ldots, m; j = 1, 2, \ldots, n), \\ \omega_j^l \leq \omega_j \leq \omega_j^u \ (j = 1, 2, \ldots, n), \\ \rho_j^l \leq \rho_j \leq \rho_j^u \ (j = 1, 2, \ldots, n), \\ \sigma_j^l \leq \sigma_j \leq \sigma_j^u \ (j = 1, 2, \ldots, n), \\ \omega_j \geq \sigma_j \qquad (j = 1, 2, \ldots, n), \\ \rho_j \geq \sigma_j \qquad (j = 1, 2, \ldots, n), \end{cases} \tag{6}$$

and

Find $\left(\mu_{ij}, \nu_{ij}, \omega_j, \rho_j, \sigma_j\right)$

so as to

$$Min\ C_i = \frac{\sum_{j=1}^{n}\left[\omega_j \mu_{ij} + \rho_j(1 - \nu_{ij}) + \sigma_j\left(1 - \mu_{ij} - \nu_{ij}\right)\right]}{\sum_{j=1}^{n}\left[\omega_j + \rho_j + 2\sigma_j\left(1 - \mu_{ij} - \nu_{ij}\right)\right]}$$

$$\text{subject to} \begin{cases} \mu_{ij}^l \leq \mu_{ij} \leq \mu_{ij}^u \ (i = 1, 2, \ldots, m; j = 1, 2, \ldots, n), \\ \nu_{ij}^l \leq \nu_{ij} \leq \nu_{ij}^u \ (i = 1, 2, \ldots, m; j = 1, 2, \ldots, n), \\ \omega_j^l \leq \omega_j \leq \omega_j^u \ (j = 1, 2, \ldots, n), \\ \rho_j^l \leq \rho_j \leq \rho_j^u \ (j = 1, 2, \ldots, n), \\ \sigma_j^l \leq \sigma_j \leq \sigma_j^u \ (j = 1, 2, \ldots, n) \\ \omega_j \geq \sigma_j \qquad (j = 1, 2, \ldots, n) \\ \rho_j \geq \sigma_j \qquad (j = 1, 2, \ldots, n) \end{cases} \tag{7}$$

As $logC_i$ and C_i possess identical monotonic behaviour, hence C_i is a monotonically increasing function with respect to μ_{ij} and monotonically decreasing function with respect to ν_{ij}.

It follows that the RCC function C_i attains maximum value at $\mu_{ij} = \mu_{ij}^u$ and $\nu_{ij} = \nu_{ij}^l$ which are, respectively, the upper bounds and lower bounds of the intervals $\left[\mu_{ij}^l, \mu_{ij}^u\right]$ and $\left[\nu_{ij}^l, \nu_{ij}^u\right]$.

Hence, the NLFP model (6) can be simplified to the following linear fractional programming model:

Find $\left(\omega_j, \rho_j, \sigma_j\right)$

so as to

$$Max\ C_i = \frac{\sum_{j=1}^{n}\left[\omega_j\mu_{ij}^u + \rho_j\left(1 - \nu_{ij}^l\right) + \sigma_j\left(1 - \mu_{ij}^u - \nu_{ij}^l\right)\right]}{\sum_{j=1}^{n}\left[\omega_j + \rho_j + 2\sigma_j\left(1 - \mu_{ij}^u - \nu_{ij}^l\right)\right]}$$

$$\text{subject to}\begin{cases} \omega_j^l \le \omega_j \le \omega_j^u\ (j=1,2,\ldots,n) \\ \rho_j^l \le \rho_j \le \rho_j^u\ (j=1,2,\ldots,n) \\ \sigma_j^l \le \sigma_j \le \sigma_j^u\ (j=1,2,\ldots,n) \\ \omega_j \ge \sigma_j \qquad (j=1,2,\ldots,n) \\ \rho_j \ge \sigma_j \qquad (j=1,2,\ldots,n) \end{cases}$$ (8)

Similarly, the NLFP model (7) can be simplified to the following linear fractional programming model:

Find $\left(\omega_j, \rho_j, \sigma_j\right)$

so as to

$$Min\ C_i = \frac{\sum_{j=1}^{n}\left[\omega_j\mu_{ij}^l + \rho_j\left(1 - \nu_{ij}^u\right) + \sigma_j\left(1 - \mu_{ij}^l - \nu_{ij}^u\right)\right]}{\sum_{j=1}^{n}\left[\omega_j + \rho_j + 2\sigma_j\left(1 - \mu_{ij}^l - \nu_{ij}^u\right)\right]}$$

$$\text{subject to}\begin{cases} \omega_j^l \le \omega_j \le \omega_j^u\ (j=1,2,\ldots,n) \\ \rho_j^l \le \rho_j \le \rho_j^u\ (j=1,2,\ldots,n) \\ \sigma_j^l \le \sigma_j \le \sigma_j^u\ (j=1,2,\ldots,n) \\ \omega_j \ge \sigma_j \qquad (j=1,2,\ldots,n) \\ \rho_j \ge \sigma_j \qquad (j=1,2,\ldots,n) \end{cases}$$ (9)

Based on variable transformation approach [18, 24, 25], the linear fractional programming models (8) and (9) are converted into equivalent linear programming models (12) and (13), respectively, using the following transformations:

$$\text{Let } z = \frac{1}{\sum_{j=1}^{n} \left[\omega_j + \rho_j + 2\sigma_j \left(1 - \mu_{ij}^u - \nu_{ij}^l \right) \right]}.$$

$$\Rightarrow \sum_{j=1}^{n} \left[z\omega_j + z\rho_j + 2z\sigma_j \left(1 - \mu_{ij}^u - \nu_{ij}^l \right) \right] = 1 \tag{10}$$

$$\Rightarrow \sum_{j=1}^{n} \left[t_j + x_j + 2y_j \left(1 - \mu_{ij}^u - \nu_{ij}^l \right) \right] = 1$$

where $z\omega_j = t_j$, $z\rho_j = x_j$ and $z\sigma_j = y_j$

Similarly, considering $u = \frac{1}{\sum_{j=1}^{n} \left[\omega_j + \rho_j + 2\sigma_j \left(1 - \mu_{ij}^l - \nu_{ij}^u \right) \right]}$, it can be found that

$$\Rightarrow \sum_{j=1}^{n} \left[a_j + b_j + 2c_j \left(1 - \mu_{ij}^l - \nu_{ij}^u \right) \right] = 1 \tag{11}$$

where $u\omega_j = a_j$, $u\rho_j = b_j$, $u\sigma_j = c_j$

Using Eq. (10), the linear fractional programming model (8) reduces to the equivalent linear programming model as follows:

Find (z, t_j, x_j, y_j)

so as to

$$Max\ C_i = \sum_{j=1}^{n} \left[t_j \mu_{ij}^u + x_j \left(1 - \nu_{ij}^l \right) + y_j \left(1 - \mu_{ij}^u - \nu_{ij}^l \right) \right]$$

$$\text{subject to} \begin{cases} z\omega_j^l \le t_j \le z\omega_j^u (j = 1, 2, \ldots, n) \\ z\rho_j^l \le x_j \le z\rho_j^u (j = 1, 2, \ldots, n) \\ z\sigma_j^l \le y_j \le z\sigma_j^u (j = 1, 2, \ldots, n) \\ \sum_{j=1}^{n} \left[t_j + x_j + 2y_j \left(1 - \mu_{ij}^u - \nu_{ij}^l \right) \right] = 1 \\ t_j \ge y_j (j = 1, 2, \ldots, n) \\ x_j \ge y_j (j = 1, 2, \ldots, n) \\ z \ge 0 \end{cases} \tag{12}$$

Similarly, using Eq. (11), the linear fractional programming model (9) reduces to the following form as

Find $\left(u, a_j, b_j, c_j\right)$

so as to

$$Min\, C_i = \sum_{j=1}^{n} \left[a_j\mu_{ij}^l + b_j\left(1 - \nu_{ij}^u\right) + c_j\left(1 - \mu_{ij}^l - \nu_{ij}^u\right)\right]$$

$$\text{subject to} \begin{cases} u\omega_j^l \le a_j \le u\omega_j^u \, (j = 1, 2, \ldots, n) \\ u\rho_j^l \le b_j \le u\rho_j^u \, (j = 1, 2, \ldots, n) \\ u\sigma_j^l \le c_j \le u\sigma_j^u \, (j = 1, 2, \ldots, n) \\ \sum_{j=1}^{n} \left[a_j + b_j + 2c_j\left(1 - \mu_{ij}^l - \nu_{ij}^u\right)\right] = 1 \\ a_j \ge c_j \, (j = 1, 2, \ldots, n) \\ b_j \ge c_j \, (j = 1, 2, \ldots, n) \\ u \ge 0 \end{cases} \quad (13)$$

Now, let $C_i^u = Max\, C_i$ and $C_i^l = Min\, C_i$ represent the upper and the lower bounds of the interval of RCCs given by models (12) and (13), respectively. Thus, the intervals of RCCs for the alternatives $A_i(i = 1, 2, \ldots, m)$ are $\left[C_i^l, C_i^u\right]$ $(i = 1, 2, \ldots, m)$.

The linear programming models (12) and (13) are solved using software LINGO (Ver. 15.0).

The order $A_i \succeq A_r$ represents "Alternative A_i is not inferior to A_r", and its degree of possibility is measured by the possibility degree of $C_i \ge C_r$. Hence, by Definition 2.3, the possibility degree of $A_i \succeq A_r$, is given by

$$p_{ir} = p(A_i \succeq A_r) = p(C_i \ge C_r) = min\left\{max\left\{\frac{C_i^u - C_r^l}{L(C_i) + L(C_r)}, 0\right\}, 1\right\} \quad (14)$$

where $C_i = \left[C_i^l, C_i^u\right], C_r = \left[C_r^l, C_r^u\right], L(C_i) = C_i^u - C_i^l$ and $L(C_r) = C_r^u - C_r^l$

Using Eq. (14), the possibility degree matrix as fuzzy preference relation obtained by pairwise comparison of the intervals of RCCs $C_i = \left[C_i^l, C_i^u\right]$ $(i = 1, 2, \ldots, m)$ takes the following form:

$$P = \left(p_{ij}\right)_{m \times m} = \begin{array}{c} \\ A_1 \\ A_2 \\ \vdots \\ A_m \end{array} \begin{array}{c} A_1 \quad A_2 \quad \cdots \quad A_m \\ \begin{pmatrix} p_{11} & p_{12} & \cdots & p_{1m} \\ p_{21} & p_{22} & \cdots & p_{2m} \\ \vdots & \vdots & \cdots & \vdots \\ p_{m1} & p_{m2} & \cdots & p_{mm} \end{pmatrix} \end{array}$$

Xu and Da [27] proposed the optimal degree formula for estimating the crisp performance scores of alternatives from the possibility degree matrix P which follows as:

$$\theta_i = \frac{1}{m(m-1)} \left(\sum_{k=1}^{n} p_{ik} + \frac{m}{2} - 1 \right), \text{for } i = 1, 2, \ldots, m \tag{15}$$

Expressing Eq. (15) as

$$\theta_i = \frac{1}{m-1} \left(\frac{1}{m} \sum_{k=1}^{n} p_{ik} \right) + \frac{m-2}{2m(m-1)} = \frac{1}{m-1} \xi_i + \frac{m-2}{2m(m-1)}, \quad \text{where} \quad \xi_i = \frac{1}{m} \sum_{k=1}^{n} p_{ik}, \text{for}$$
$i = 1, 2, \ldots, m$

Thus, the above relation takes the functional form

$$\theta = \frac{1}{m-1} \xi + \frac{m-2}{2m(m-1)}$$

$\Rightarrow \frac{d\theta}{d\xi} = \frac{1}{m-1} > 0$ (Since, the no. of alternatives $m \geq 2$ remains same in a given MADM problem).

This implies that θ is strictly increasing function of ξ, and hence for ease of calculation, the ranking of alternatives $A_i (i = 1, 2, \ldots, m)$ can be effectively done using the relation

$$\xi_i = \frac{1}{m} \sum_{k=1}^{n} p_{ik}, \text{for } i = 1, 2, \ldots, m \tag{16}$$

The decreasing order of values of ξ_i provides the ranking order of alternatives $A_i (i = 1, 2, \ldots, m)$.

4 The Algorithm

Based on previous deductions, the TOPSIS for solving the MADM problems in IVIF settings are described through the following algorithmic form:

Step 1: In a given MADM problem, n criteria $\Gamma_j (j = 1, 2, \ldots, n)$ and m alternatives $A_i (i = 1, 2, \ldots, m)$ are identified.

Step 2: The alternatives $A_i (i = 1, 2, \ldots, m)$ are evaluated by a decision maker relative to criteria $\Gamma_j (j = 1, 2, \ldots, n)$ generating the three parameter IVIF decision matrix as

$$D = (\alpha_{ij})_{m \times n} = \left(\langle \left[\mu_{ij}^l, \mu_{ij}^u \right], \left[\nu_{ij}^l, \nu_{ij}^u \right], \left[\pi_{ij}^l, \pi_{ij}^u \right] \rangle \right)_{m \times n}.$$

Step 3: Decision maker's opinions on criteria $\Gamma_j (j=1,2,\ldots,n)$ generate the three parameter IVIF criteria weight vector $w = (w_j)_{1 \times n} = \left(\left[\omega_j^l, \omega_j^u \right], \left[\rho_j^l, \rho_j^u \right], \left[\sigma_j^l, \sigma_j^u \right] \right)_{1 \times n}$.

Step 4: Using models (12) and (13), linear programming models are constructed for finding upper and lower bounds of intervals of RCCs $C_i (i=1,2,\ldots m)$ for alternatives $A_i (i=1,2,\ldots,m)$ under TOPSIS approach.

Step 5: The intervals of RCCs $C_i = \left[C_i^l, C_i^u \right]$ for alternatives $A_i (i=1,2,\ldots,m)$ are generated.

Step 6: Using Eq. (14), the fuzzy preference relation as possibility degree matrix $P = (p_{ij})_{m \times m}$ is obtained by pairwise comparison of intervals of RCCs C_i of alternatives $A_i (i=1,2,\ldots,m)$.

Step 7: By applying Eq. (16) on the fuzzy preference relation $P = (p_{ij})_{m \times m}$, the optimal degrees ξ_i of alternatives $A_i (i=1,2,\ldots,m)$ are estimated.

Step 8: The priority scores λ_i of alternatives $A_i (i=1,2,\ldots,m)$ are obtained by normalizing the optimal degrees $\xi_i (i=1,2,\ldots,m)$.

Step 9: The alternatives $A_i (i=1,2,\ldots,m)$ are ranked in decreasing order of the priority scores λ_i.

5   Numerical Illustrations

In this section, two MADM problems have been considered [19, 24, 29] for illustrating the application of the proposed TOPSIS approach in IVIF environment.

5.1   Illustrative Example 1

The problem is related to an investment company planning to invest in the four companies (alternatives) A_1 (car company), A_2 (food company), A_3 (computer company) and A_4 (arms company). These companies are evaluated relative to three attributes Γ_1 (risk analysis), Γ_2 (growth analysis) and Γ_3 (environmental impact analysis) by an expert. The IVIF decision matrix obtained in the process is expressed as

$$D = (a_{ij})_{4 \times 3} = (([\mu_{ij}^l, \mu_{ij}^u], [v_{ij}^l, v_{ij}^u], [\pi_{ij}^l, \pi_{ij}^u]))_{4 \times 3} =$$

$$
\begin{array}{cccc}
 & \Gamma_1 & \Gamma_2 & \Gamma_3 \\
A_1 & \langle [0.4,0.5],[0.3,0.4],[0.1,0.3] \rangle & \langle [0.4,0.6],[0.2,0.4],[0.0,0.4] \rangle & \langle [0.1,0.3],[0.5,0.6],[0.1,0.4] \rangle \\
A_2 & \langle [0.6,0.7],[0.2,0.3],[0.0,0.2] \rangle & \langle [0.6,0.7],[0.2,0.3],[0.0,0.2] \rangle & \langle [0.4,0.7],[0.1,0.2],[0.1,0.5] \rangle \\
A_3 & \langle [0.3,0.6],[0.3,0.4],[0.0,0.4] \rangle & \langle [0.5,0.6],[0.3,0.4],[0.0,0.2] \rangle & \langle [0.5,0.6],[0.1,0.3],[0.1,0.4] \rangle \\
A_4 & \langle [0.7,0.8],[0.1,0.2],[0.0,0.2] \rangle & \langle [0.6,0.7],[0.1,0.3],[0.0,0.3] \rangle & \langle [0.3,0.4],[0.1,0.2],[0.4,0.6] \rangle
\end{array}
$$

The IVIF criteria weight vector is expressed as

$$W = (w_1, w_2, w_3) = (\langle[0.1, 0.4], [0.2, 0.55], [0.05, 0.7]\rangle, \langle[0.2, 0.5], [0.15, 0.45], [0.05, 0.65]\rangle,$$
$$\langle[0.25, 0.6], [0.15, 0.38], [0.02, 0.6]\rangle)$$

Using linear programming (LP) models (12) and (13), the auxiliary linear programming problems (LPPs) for alternative A_1 are given as follows

$$max\ C_1 = 0.5t_1 + 0.6t_2 + 0.3t_3 + 0.7x_1 + 0.8x_2 + 0.5x_3 + 0.2y_1 + 0.2y_2 + 0.2y_3$$

$$\text{subject to} \begin{cases} 0.1z \leq t_1 \leq 0.4z, \\ 0.2z \leq t_2 \leq 0.5z, \\ 0.25z \leq t_3 \leq 0.6z, \\ 0.2z \leq x_1 \leq 0.55z, \\ 0.15z \leq x_2 \leq 0.45z, \\ 0.15z \leq x_3 \leq 0.38z, \\ 0.05z \leq y_1 \leq 0.7z, \\ 0.05z \leq y_2 \leq 0.65z, \\ 0.02z \leq y_3 \leq 0.6z, \\ t_1 + t_2 + t_3 + x_1 + x_2 + x_3 + 0.4y_1 + 0.4y_2 + 0.4y_3 = 1, \\ t_j \geq y_j (j = 1, 2, 3), \\ x_j \geq y_j (j = 1, 2, 3), \\ z \geq 0 \end{cases}$$

$$(17)$$

$$min\ C_1 = 0.4a_1 + 0.4a_2 + 0.1a_3 + 0.6b_1 + 0.6b_2 + 0.4b_3 + 0.2c_1 + 0.2c_2 + 0.3c_3$$

$$\text{subject to} \begin{cases} 0.1u \leq a_1 \leq 0.4u, \\ 0.2u \leq a_2 \leq 0.5u, \\ 0.25u \leq a_3 \leq 0.6u, \\ 0.2u \leq b_1 \leq 0.55u, \\ 0.15u \leq b_2 \leq 0.45u, \\ 0.15u \leq b_3 \leq 0.38u, \\ 0.05u \leq c_1 \leq 0.7u, \\ 0.05u \leq c_2 \leq 0.65u, \\ 0.02u \leq c_3 \leq 0.6u, \\ a_1 + a_2 + a_3 + b_1 + b_2 + b_3 + 0.4c_1 + 0.4c_2 + 0.6c_3 = 1, \\ a_j \geq c_j (j = 1, 2, 3), \\ b_j \geq c_j (j = 1, 2, 3), \\ u \geq 0 \end{cases}$$

$$(18)$$

The optimal values of the objective functions of LPPs (17) and (18) are estimated using LINGO (Ver. 15.0) software as $C_1^u = 0.623$ and $C_1^l = 0.328$, respectively. Hence, the interval of relative closeness coefficient for alternative A_1 is given by $C_1 = [0.328, 0.623]$.

Using LP models (12) and (13), the auxiliary LPPs for alternative A_2 are given by

$$max\ C_2 = 0.7t_1 + 0.7t_2 + 0.7t_3 + 0.8x_1 + 0.8x_2 + 0.9x_3 + 0.1y_1 + 0.1y_2 + 0.2y_3$$

$$\text{subject to} \begin{cases} 0.1z \leq t_1 \leq 0.4z, \\ 0.2z \leq t_2 \leq 0.5z, \\ 0.25z \leq t_3 \leq 0.6z, \\ 0.2z \leq x_1 \leq 0.55z, \\ 0.15z \leq x_2 \leq 0.45z, \\ 0.15z \leq x_3 \leq 0.38z, \\ 0.05z \leq y_1 \leq 0.7z, \\ 0.05z \leq y_2 \leq 0.65z, \\ 0.02z \leq y_3 \leq 0.6z, \\ t_1 + t_2 + t_3 + x_1 + x_2 + x_3 + 0.2y_1 + 0.2y_2 + 0.4y_3 = 1, \\ t_j \geq y_j (j = 1, 2, 3), \\ x_j \geq y_j (j = 1, 2, 3), \\ z \geq 0 \end{cases}$$

(19)

$$min\ C_2 = 0.6a_1 + 0.6a_2 + 0.4a_3 + 0.7b_1 + 0.7b_2 + 0.8b_3 + 0.1c_1 + 0.1c_2 + 0.4c_3$$

$$\text{subject to} \begin{cases} 0.1u \leq a_1 \leq 0.4u, \\ 0.2u \leq a_2 \leq 0.5u, \\ 0.25u \leq a_3 \leq 0.6u, \\ 0.2u \leq b_1 \leq 0.55u, \\ 0.15u \leq b_2 \leq 0.45u, \\ 0.15u \leq b_3 \leq 0.38u, \\ 0.05u \leq c_1 \leq 0.7u, \\ 0.05u \leq c_2 \leq 0.65u, \\ 0.02u \leq c_3 \leq 0.6u, \\ a_1 + a_2 + a_3 + b_1 + b_2 + b_3 + 0.2c_1 + 0.2c_2 + 0.8c_3 = 1, \\ a_j \geq c_j (j = 1, 2, 3), \\ b_j \geq c_j (j = 1, 2, 3), \\ u \geq 0 \end{cases}$$

(20)

The optimal values of the objective functions of LPPs (19) and (20) are estimated using LINGO (Ver. 15.0) software as $C_2^u = 0.787$ and $C_2^l = 0.554$, respectively. Hence, the interval of RCC for alternative A_2 is given by $C_2 = [0.554, 0.787]$.

The auxiliary LPPs for alternative A_3 using LP models (12) and (13) are given by

$$max\ C_3 = 0.6t_1 + 0.6t_2 + 0.6t_3 + 0.7x_1 + 0.7x_2 + 0.9x_3 + 0.1y_1 + 0.1y_2 + 0.3y_3$$

$$\text{subject to} \begin{cases} 0.1z \leq t_1 \leq 0.4z, \\ 0.2z \leq t_2 \leq 0.5z, \\ 0.25z \leq t_3 \leq 0.6z, \\ 0.2z \leq x_1 \leq 0.55z, \\ 0.15z \leq x_2 \leq 0.45z, \\ 0.15z \leq x_3 \leq 0.38z, \\ 0.05z \leq y_1 \leq 0.7z, \\ 0.05z \leq y_2 \leq 0.65z, \\ 0.02z \leq y_3 \leq 0.6z, \\ t_1 + t_2 + t_3 + x_1 + x_2 + x_3 + 0.2y_1 + 0.2y_2 + 0.6y_3 = 1, \\ t_j \geq y_j (j = 1, 2, 3), \\ x_j \geq y_j (j = 1, 2, 3), \\ z \geq 0 \end{cases}$$

$$\text{(21)}$$

$$min\ C_3 = 0.3a_1 + 0.5a_2 + 0.5a_3 + 0.6b_1 + 0.6b_2 + 0.7b_3 + 0.3c_1 + 0.1c_2 + 0.2c_3$$

$$\text{subject to} \begin{cases} 0.1u \leq a_1 \leq 0.4u, \\ 0.2u \leq a_2 \leq 0.5u, \\ 0.25u \leq a_3 \leq 0.6u, \\ 0.2u \leq b_1 \leq 0.55u, \\ 0.15u \leq b_2 \leq 0.45u, \\ 0.15u \leq b_3 \leq 0.38u, \\ 0.05u \leq c_1 \leq 0.7u, \\ 0.05u \leq c_2 \leq 0.65u, \\ 0.02u \leq c_3 \leq 0.6u, \\ a_1 + a_2 + a_3 + b_1 + b_2 + b_3 + 0.6c_1 + 0.2c_2 + 0.4c_3 = 1, \\ a_j \geq c_j (j = 1, 2, 3), \\ b_j \geq c_j (j = 1, 2, 3), \\ u \geq 0 \end{cases}$$

$$\text{(22)}$$

Using LINGO (Ver. 15.0) software, the optimal values of the objective functions of LPPs (21) and (22) are estimated as $C_3^u = 0.711$ and $C_3^l = 0.489$, respectively. Hence, the interval of RCC for alternative A_3 is given by $C_3 = [0.489, 0.711]$.

The auxiliary LPPs for alternative A_4 using LP models (12) and (13) are given by

$$max\ C_4 = 0.8t_1 + 0.7t_2 + 0.4t_3 + 0.9x_1 + 0.9x_2 + 0.9x_3 + 0.1y_1 + 0.2y_2 + 0.5y_3$$

$$\text{subject to} \begin{cases} 0.1z \leq t_1 \leq 0.4z, \\ 0.2z \leq t_2 \leq 0.5z, \\ 0.25z \leq t_3 \leq 0.6z, \\ 0.2z \leq x_1 \leq 0.55z, \\ 0.15z \leq x_2 \leq 0.45z, \\ 0.15z \leq x_3 \leq 0.38z, \\ 0.05z \leq y_1 \leq 0.7z, \\ 0.05z \leq y_2 \leq 0.65z, \\ 0.02z \leq y_3 \leq 0.6z, \\ t_1 + t_2 + t_3 + x_1 + x_2 + x_3 + 0.2y_1 + 0.4y_2 + y_3 = 1, \\ t_j \geq y_j (j = 1, 2, 3), \\ x_j \geq y_j (j = 1, 2, 3), \\ z \geq 0 \end{cases}$$

$$(23)$$

$$min\ C_4 = 0.7a_1 + 0.6a_2 + 0.3a_3 + 0.8b_1 + 0.7b_2 + 0.8b_3 + 0.1c_1 + 0.1c_2 + 0.5c_3$$

$$\text{subject to} \begin{cases} 0.1u \leq a_1 \leq 0.4u, \\ 0.2u \leq a_2 \leq 0.5u, \\ 0.25u \leq a_3 \leq 0.6u, \\ 0.2u \leq b_1 \leq 0.55u, \\ 0.15u \leq b_2 \leq 0.45u, \\ 0.15u \leq b_3 \leq 0.38u, \\ 0.05u \leq c_1 \leq 0.7u, \\ 0.05u \leq c_2 \leq 0.65u, \\ 0.02u \leq c_3 \leq 0.6u, \\ a_1 + a_2 + a_3 + b_1 + b_2 + b_3 + 0.2c_1 + 0.2c_2 + c_3 = 1, \\ a_j \geq c_j (j = 1, 2, 3), \\ b_j \geq c_j (j = 1, 2, 3), \\ u \geq 0 \end{cases}$$

$$(24)$$

Using LINGO (Ver. 15.0) software, the optimal values of the objective functions of LPPs (23) and (24) are computed as $C_4^u = 0.802$ and $C_4^l = 0.534$, respectively. Hence, the interval of RCC for alternative A_4 is given by $C_4 = [0.534, 0.802]$.

Thus, it is found that the comprehensive evaluation of alternatives A_1, A_2, A_3 and A_4 yields intervals of closeness coefficients as follows:

$$C_1 = [C_1^l, C_1^u] = [0.328, 0.623], C_2 = [C_2^l, C_2^u] = [0.553, 0.787]$$
$$C_3 = [C_3^l, C_3^u] = [0.489, 0.711], C_4 = [C_4^l, C_4^u] = [0.534, 0.802]$$

Using Eq. (14), the fuzzy preference relation as possibility degree matrix for pairwise comparison of alternatives $A_i (i = 1, 2, 3, 4)$ is given by

$$P = (p_{ij})_{4 \times 4} = \begin{array}{c} \\ A_1 \\ A_2 \\ A_3 \\ A_4 \end{array} \begin{array}{cccc} A_1 & A_2 & A_3 & A_4 \\ \left(\begin{array}{cccc} 0.5 & 0.132 & 0.259 & 0.158 \\ 0.868 & 0.5 & 0.654 & 0.504 \\ 0.741 & 0.346 & 0.5 & 0.361 \\ 0.842 & 0.496 & 0.639 & 0.5 \end{array} \right) \end{array}$$

Using Eq. (16), optimal degrees ξ_i of alternatives $A_i (i = 1, 2, 3, 4)$ are given by

$$\xi_1 = 0.262, \xi_2 = 0.632, \xi_3 = 0.487, \xi_4 = 0.619$$

The priority scores λ_i for alternatives $A_i (i = 1, 2, 3, 4)$ are obtained by normalizing the optimal degrees $\xi_i (i = 1, 2, 3, 4)$. It is found that

$$\lambda_1 = 0.131, \lambda_2 = 0.316, \lambda_3 = 0.244, \lambda_4 = 0.309$$

Thus, the ranking of alternatives becomes $A_2 \succ A_4 \succ A_3 \succ A_1$.

5.2   Illustrative Example 2

The same problem as described in illustrative example 1 is considered with the modifications that the elements of the IVIF decision matrix are with increasing uncertainties in membership and non-membership degrees by randomly increasing the width of these intervals. The revised IVIF decision matrix is given by

$$D = (a_{ij})_{4 \times 3} = \left(\langle \left[\mu_{ij}^l, \mu_{ij}^u \right], \left[\nu_{ij}^l, \nu_{ij}^u \right], \left[\pi_{ij}^l, \pi_{ij}^u \right] \rangle \right)_{4 \times 3} =$$

	Γ_1	Γ_2	Γ_3
A_1	$\langle [0.3, 0.7], [0.1, 0.3], [0.0, 0.6] \rangle$	$\langle [0.3, 0.6], [0.1, 0.4], [0.0, 0.6] \rangle$	$\langle [0.1, 0.6], [0.1, 0.4], [0.0, 0.8] \rangle$
A_2	$\langle [0.4, 0.7], [0.1, 0.3], [0.0, 0.5] \rangle$	$\langle [0.3, 0.6], [0.2, 0.3], [0.1, 0.5] \rangle$	$\langle [0.4, 0.7], [0.1, 0.2], [0.1, 0.5] \rangle$
A_3	$\langle [0.3, 0.6], [0.3, 0.4], [0.0, 0.4] \rangle$	$\langle [0.2, 0.6], [0.2, 0.3], [0.1, 0.6] \rangle$	$\langle [0.3, 0.7], [0.1, 0.3], [0.0, 0.6] \rangle$
A_4	$\langle [0.4, 0.7], [0.1, 0.2], [0.1, 0.5] \rangle$	$\langle [0.3, 0.7], [0.1, 0.3], [0.0, 0.6] \rangle$	$\langle [0.3, 0.6], [0.1, 0.2], [0.2, 0.6] \rangle$

The IVIF criteria weight vector remains same as

$$W = (w_1, w_2, w_3) = (\langle [0.1, 0.4], [0.2, 0.55], [0.05, 0.7] \rangle, \langle [0.2, 0.5], [0.15, 0.45], [0.05, 0.65] \rangle,$$
$$\langle [0.25, 0.6], [0.15, 0.38], [0.02, 0.6] \rangle)$$

This problem is solved by following the linear programming based TOPSIS procedures as adopted in solving illustrative example 1. The interval performance scores of alternatives A_1, A_2, A_3 and A_4 as intervals of RCCs are as follows:

$$C_1 = [0.3325, 0.8097], C_2 = [0.4593, 0.8021], C_3 = [0.3712, 0.7515], C_4 = [0.4405, 0.8214]$$

Using Eq. (14), the fuzzy preference relation as possibility degree matrix for pairwise comparison of intervals of RCCs for alternatives $A_i (i = 1, 2, 3, 4)$ is given by

$$P = (p_{ij})_{4 \times 4} = \begin{array}{c} \\ A_1 \\ A_2 \\ A_3 \\ A_4 \end{array} \begin{pmatrix} A_1 & A_2 & A_3 & A_4 \\ 0.5 & 0.427 & 0.511 & 0.430 \\ 0.573 & 0.5 & 0.596 & 0.500 \\ 0.489 & 0.404 & 0.5 & 0.409 \\ 0.570 & 0.500 & 0.591 & 0.5 \end{pmatrix}$$

The priority scores λ_i for alternatives $A_i (i = 1, 2, 3, 4)$ are obtained as

$$\lambda_1 = 0.234, \lambda_2 = 0.271, \lambda_3 = 0.225, \lambda_4 = 0.270$$

Thus, the ranking of alternatives becomes $A_2 \succ A_4 \succ A_1 \succ A_3$.

Following the approach of Li [24], the interval performance scores of alternatives A_1, A_2, A_3 and A_4 as intervals of RCCs are given by

$$C_1 = [0.325, 0.8197], C_2 = [0.4575, 0.811], C_3 = [0.365, 0.7576],$$
$$C_4 = [0.4375, 0.8301]$$

By Eq. (14), the fuzzy preference relation as possibility degree matrix for pairwise comparison of intervals of RCCs for alternatives $A_i (i = 1, 2, 3, 4)$ is given by

$$P = (p_{ij})_{4 \times 4} = \begin{array}{c} \\ A_1 \\ A_2 \\ A_3 \\ A_4 \end{array} \begin{pmatrix} A_1 & A_2 & A_3 & A_4 \\ 0.5 & 0.427 & 0.5125 & 0.4307 \\ 0.573 & 0.5 & 0.5978 & 0.5001 \\ 0.4875 & 0.4022 & 0.5 & 0.4077 \\ 0.5693 & 0.4999 & 0.5923 & 0.5 \end{pmatrix}$$

The performance scores θ_i for alternatives $A_i (i = 1, 2, 3, 4)$ are obtained as

$$\theta_1 = 0.239, \theta_2 = 0.264, \theta_3 = 0.233, \theta_4 = 0.263$$

Thus, the ranking of alternatives becomes $A_2 \succ A_4 \succ A_1 \succ A_3$.

The results obtained in the previous two illustrative examples are summarized in Table 1 and comparatively analysed in Table 2.

5.3   Results and Discussions

- In illustrative example 1, it is found that from Table 1 that the ranking of alternatives $A_2 \succ A_4 \succ A_3 \succ A_1$ based on the proposed approach becomes

Table 1 Comparison of the proposed approach with other approaches

Example	Approach	Criteria weights	Criteria	Ranking
Illustrative example 1	Li [24]	IVIFN	$\Gamma_1, \Gamma_2, \Gamma_3$	$A_2 \succ A_4 \succ A_3 \succ A_1$
	Ye [29]	Crisp	$\Gamma_1, \Gamma_2, \Gamma_3$	$A_2 \succ A_4 \succ A_3 \succ A_1$
	Proposed approach	IVIFN	$\Gamma_1, \Gamma_2, \Gamma_3$	$A_2 \succ A_4 \succ A_3 \succ A_1$
Illustrative example 2	Li [24]	IVIFN	$\Gamma_1, \Gamma_2, \Gamma_3$	$A_2 \succ A_4 \succ A_1 \succ A_3$
	Proposed approach	IVIFN	$\Gamma_1, \Gamma_2, \Gamma_3$	$A_2 \succ A_4 \succ A_1 \succ A_3$

identical with the rankings of both Li [24] (IVIFN criteria weights) and Ye [29] (crisp criteria weights) validating our approach.

- In illustrative example 2, it is again found from Table 1 that the ranking of alternatives $A_2 \succ A_4 \succ A_1 \succ A_3$ based on the proposed approach coincides with the ranking of Li [24] validating the proposed approach.

- The numerical results obtained in the two illustrative examples are entered in Table 2. The widths of the intervals of RCCs under Li [24] and proposed approach for illustrative example 1 are compared and observed that except for the alternative A_2, the widths of the intervals of RCCs by the proposed approach is smaller than those by the Li's approach. Lesser widths of C_i's imply lesser uncertainty in the selection of alternatives. Thus, though all the rankings coincide, the ordering of alternatives by the proposed approach involves less uncertainty.

- The illustrative example 2 is similar to the illustrative example 1 with the only difference in the IVIF decision matrix which is generated by increasing uncertainties in membership and non-membership interval degrees by randomly increasing the width of these intervals. The widths of the intervals of RCCs under Li [24] and proposed approach for illustrative example 2 are compared and observed that the widths of all the intervals of RCCs C_i $(i = 1, 2, 3, 4)$ under the proposed approach is smaller than those under the Li's approach. Further, the RCC intervals of alternatives for illustrative example 2 under the proposed approach are entirely contained in the respective RCC intervals under Li's approach [24]. Thus, from these observations, it may be concluded that under greater uncertainties in decision matrix, the proposed approach gives better RCC interval estimates of alternatives with lesser uncertainty by shortening the widths of such intervals than Li [24].

- Some more advanced approaches, like intuitionistic fuzzy possibility degree measures at the place of possibility degree measures, may be used for deriving ranking from the intervals of RCCs for reflecting the impact of lesser uncertainties in such intervals under the proposed approach.

Table 2 Comparative analysis of the proposed approach with the Li's approach [24]

Example	Alternatives	Li's approach [24]			Proposed approach		
		Intervals of RCCs (C_i)	Width $L(C_i)$	Performance scores (θ_i)	Intervals of RCCs (C_i)	Width $L(C_i)$	Priority scores (λ_i)
Illustrative example 1	A_1	[0.321, 0.626]	0.305	0.125	[0.328, 0.623]	0.295	0.131
	A_2	[0.561, 0.791]	0.230	0.348	[0.554, 0.787]	0.233	0.316
	A_3	[0.489, 0.716]	0.227	0.261	[0.489, 0.711]	0.222	0.244
	A_4	[0.539, 0.809]	0.270	0.286	[0.534, 0.802]	0.268	0.309
Illustrative example 2	A_1	[0.325, 0.820]	0.495	0.239	[0.332, 0.810]	0.478	0.234
	A_2	[0.458, 0.811]	0.353	0.264	[0.459, 0.802]	0.343	0.271
	A_3	[0.365, 0.758]	0.393	0.233	[0.371, 0.752]	0.381	0.225
	A_4	[0.437, 0.830]	0.393	0.263	[0.440, 0.821]	0.381	0.270

- The weighted absolute distance between IFSs with three parameter characterizations of IFSs is used in proposed LPP-based TOPSIS approach in IVIF environment. The computational complexities of the proposed approach are increased due to the presence of the third parameter intuitionistic index in the formulation.

6 Concluding Remarks

Szmidt and Kacprzyk [26] established that the third parameter, hesitancy degree, has influence on distance measure in IF settings and proved that any extension of fuzzy distance measure in the IF environment should contain all the three parameters: membership, non-membership and hesitancy degrees. Xu and Yager [28] also defined distance between IFNs with three-parameter characterizations of IFNs. Thus, a weighted absolute distance between two IFSs with respect to IF weights is defined as an extension of the weighted absolute distance [24]. Fractional programming technique-based TOPSIS approach in IVIF environment generates a pair of nonlinear fractional programming models that are transformed into two auxiliary linear programming models for solving MADM problems resulting in intervals of RCCs of alternatives. The fuzzy preference relation as possibility degree matrix for pairwise comparisons of RCC intervals of alternatives is generated, and priority scores of alternatives are evaluated for ranking. It is found that the ranking of alternatives resulted from the proposed approach remains identical with the existing approaches [24, 29] validating the proposed approach. Based on the results of the two illustrative examples, it is found that under greater uncertainties in membership and non-membership interval degrees, the advanced approach gives lesser uncertainties in RCC interval estimates through smaller widths of such intervals in comparison with those under Li's approach [24]. This implies that the proposed method has the power to model imprecision and uncertainty of real-life situations in more reliable manner. The increase in computational complexities is due to the presence of hesitancy degree in RCC formulation under TOPSIS. The impact of lesser uncertainties in RCC intervals on the ranking of alternatives needs further studies by employing IF possibility degree measures at the place of possibility degree measures or some other available techniques. Some more examples may also exist demonstrating the differences in rankings under the proposed approach and other approaches. The three parameter TOPSIS approach in IVIF settings can further be simplified by future researchers for better modelling of imprecision.

Acknowledgements The authors remain grateful to the anonymous reviewers for their valuable comments and suggestions in improving the quality of the manuscript.

References

1. Atanassov, K.T.: Intuitionistic fuzzy sets. Fuzzy Sets Syst. **20**(1), 87–96 (1986)
2. Atanassov, K.T., Gargov, G.: Interval-valued intuitionistic fuzzy sets. Fuzzy Sets Syst. **31**(3), 343–349 (1989)
3. Atanassov, K.T., Pasi, G., Yager, R.R.: Intuitionistic fuzzy interpretations of multi-criteria multi-person and multi-measurement tool decision making. Int. J. Syst. Sci. **36**(4), 859–868 (2005)
4. Biswas, A., Kumar, S.: An integrated TOPSIS approach to MADM with interval-valued intuitionistic fuzzy settings. Adv. Intell. Syst. Comput. (Accepted)
5. Biswas, A., Adan, A., Halder, P., Majumdar, D., Natale, V., Randler, C., Tonetti, L., Sahu, S.: Exploration of transcultural properties of the reduced version of the morningness-eveningness questionnaire (rMEQ) using adaptive neuro fuzzy inference system. Biol. Rhythm Res. **45**, 955–968 (2014)
6. Biswas, A., Dewan, S.: Priority based fuzzy goal programming technique for solving fractional fuzzy goals by using dynamic programming. Fuzzy Inf. Eng. **4**, 165–180 (2012)
7. Biswas, A., Majumdar, D., Sahu, S.: Assessing morningness of a group of people by using fuzzy expert system and adaptive neuro fuzzy inference model. Commun. Comput. Inf. Sci. **140**, 47–56 (2011)
8. Biswas, A., Modak, N.: A fuzzy goal programming technique for multiobjective chance constrained programming with normally distributed fuzzy random variables and fuzzy numbers. Int. J. Math. Oper. Res. **5**, 551–570 (2013)
9. Biswas, A., Modak, N.: Using fuzzy goal programming technique to solve multiobjective chance constrained programming problems in a fuzzy environment. Int. J. Fuzzy Syst. Appl. **2**, 71–80 (2012)
10. Bustince, H., Burillo, P.: Vague sets are intuitionistic fuzzy sets. Fuzzy Sets Syst. **79**(3), 403–405 (1996)
11. Chen, C.T.: Extension of the TOPSIS for group decision-making under fuzzy environment. Fuzzy Sets Syst. **114**(1), 1–9 (2000)
12. Chen, S.M., Tan, J.M.: Handling multicriteria fuzzy decision-making problems based on vague set theory. Fuzzy Sets Syst. **67**(2), 163–172 (1994)
13. Chen, T.Y.: The inclusion-based TOPSIS method with interval-valued intuitionistic fuzzy sets for multiple criteria group decision making. Appl. Soft Comput. **26**, 57–73 (2015)
14. Deschrijver, G., Kerre, E.E.: On the relationship between some extensions of fuzzy set theory. Fuzzy Sets Syst. **133**(2), 227–235 (2003)
15. Deschrijver, G., Kerre, E.E.: On the position of intuitionistic fuzzy set theory in the framework of theories modeling imprecision. Inf. Sci. **177**(8), 1860–1866 (2007)
16. Facchinetti, G., Ricci, R.G., Muzzioli, S.: Note on ranking fuzzy triangular numbers. Int. J. Intell. Syst. **13**, 613–622 (1998)
17. Gau, W.L., Buehrer, D.J.: Vague sets. IEEE Trans. Syst. Man Cybern. **23**(2), 610–614 (1993)
18. Guh, Y.Y., Hon, C.C., Lee, E.S.: Fuzzy weighted average: the linear programming approach via Charnes and Cooper's rule. Fuzzy Sets Syst. **117**(1), 157–160 (2001)
19. Herrera, F., Herrera-Viedma, E.: Linguistic decision analysis: steps for solving decision problems under linguistic information. Fuzzy Sets Syst. **115**(1), 67–82 (2000)
20. Hong, D.H., Choi, C.H.: Multicriteria fuzzy decision-making problems based on vague set theory. Fuzzy Sets Syst. **114**(1), 103–113 (2000)
21. Hwang, C.L., Yoon, Y.: A state of the art survey. Multiple attribute decision making: methods and applications. Springer, Berlin (1981)
22. Klir, G.J., Yuan, B.: Fuzzy sets and fuzzy logic: theory and applications. Pearson education Inc., New Jersey, USA (1995)
23. Li, D.F., Wang, Y.C., Liu, S., Shan, F.: Fractional programming methodology for multi-attribute group decision-making using IFS. Appl. Soft Comput. **9**(1), 219–225 (2009)

24. Li, D.F.: Linear programming method for MADM with interval-valued intuitionistic fuzzy sets. Expert Syst. Appl. **37**, 5939–5945 (2010)
25. Pal, B.B., Moitra, B.N., Maulik, U.: A goal programming procedure for fuzzy multiobjective linear fractional programming problem. Fuzzy Sets Syst. **139**, 395–405 (2003)
26. Szmidt, E., Kacprzyk, J.: Distances between intuitionistic fuzzy sets. Fuzzy Sets Syst. **114**(3), 505–518 (2000)
27. Xu, Z.S., Da, Q.L.: A possibility based method for priorities of interval judgment matrices. Chin. J. Manag. Sci. **11**, 63–65 (2003)
28. Xu, Z., Yager, R.R.: Dynamic intuitionistic fuzzy multi-attribute decision making. Int. J. Approximate Reasoning **48**(1), 246–262 (2008)
29. Ye, J.: Multicriteria fuzzy decision-making method based on a novel accuracy function under interval-valued intuitionistic fuzzy environment. Expert Syst. Appl. **36**, 6899–6902 (2009)
30. Zadeh, L.A.: Fuzzy sets. Inf. Control **8**, 338–353 (1965)
31. Zeng, S., Xiao, Y.: TOPSIS method for intuitionistic fuzzy multiple-criteria decision making and its application to investment selection. Kybernetes **45**(2), 282–296 (2016)
32. Zhao, X.: TOPSIS method for interval-valued intuitionistic fuzzy multiple attribute decision making and its application to teaching quality evaluation. J. Intell. Fuzzy Syst. **26**(6), 3049–3055 (2014)

A Comparative Study of Bio-inspired Algorithms for Medical Image Registration

D. R. Sarvamangala and Raghavendra V. Kulkarni

Abstract The challenge of determining optimal transformation parameters for image registration has been treated traditionally as a multidimensional optimization problem. Non-rigid registration of medical images has been approached in this article using the particle swarm optimization algorithm, dragonfly algorithm, and the artificial bee colony algorithm. Brief introductions to these algorithms have been presented. Results of MATLAB simulations of medical image registration approached through these algorithms have been analyzed. The simulation shows that the dragonfly algorithm results in higher quality image registration, but takes longer to converge. The trade-off issue between the quality of registration and the computing time has been brought forward. This has a strong impact on the choice of the most suitable algorithm for medical applications, such as monitoring of tumor progression.

Keywords Artificial bee colony algorithm · Dragonfly algorithm
Medical image registration · Particle swarm optimization algorithm · Swarm intelligence

1 Introduction

The objective of image registration is to align structures or regions accurately across multiple, related images acquired under different times, or at different conditions, or using different modalities [1]. This important challenge in medical image processing is an active area of research. Image registration has a multitude of applications in medical image processing. Image registration is essential for image-guided surgery [2, 3], image-guided intervention [4], radiotherapy planning, cardiac perfusion [5],

D. R. Sarvamangala (✉)
REVA University, Bengaluru, India
e-mail: sarvamangaladr@reva.edu.in

R. V. Kulkarni
M. S. Ramaiah University of Applied Sciences, Bengaluru, India
e-mail: arvie@ieee.org

© Springer Nature Singapore Pte Ltd. 2019
J. K. Mandal et al. (eds.), *Advances in Intelligent Computing*,
Studies in Computational Intelligence 687,
https://doi.org/10.1007/978-981-10-8974-9_2

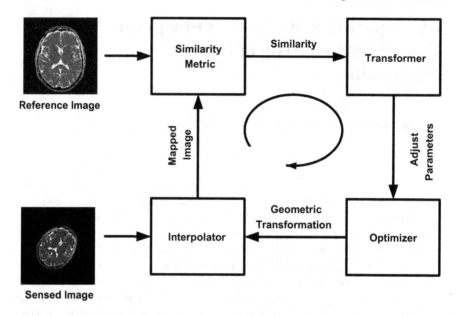

Fig. 1 The process of image registration

and monitoring of disease progression. The analysis of efficiency of treatments, such as radiotherapy and chemotherapy, requires the registration of pre-treatment and post-treatment images taken from scans. In addition, structural and functional information obtained from different imaging modalities needs to be combined for efficient determination of abnormalities which requires accurate registration. Medical image registration may involve any number of related images.

This article focuses on medical image registration using bio-inspired algorithms. In image registration, the first medical image is termed as reference image and the second as the sensed image. The procedure involves transforming the sensed image using the parameters obtained by the optimizer and calculating the similarity between the reference image and sensed image. This process is repeated until the two images are perfectly aligned. This process has been illustrated in Fig. 1.

1.1 Image Registration

Image registration is used to determine the transformation between the images. The images are either two- or three-dimensional; therefore, the transformation can be from 2-D to 2-D, from 2-D to 3-D, or from 3-D to 3-D. Intensity-based methods are widely used for registering images having same or different number of dimensions, rigid or non-rigid transformations and deformable transformations, same modality or different modality images. The reference image I_1 is fixed and does not undergo transformation, whereas the sensed image I_2 undergoes series of transformations denoted by T until it matches with the reference image. The transformed image is $I'_2 = T(I_2)$. The objective of image registration is to determine the transformation T that results in maximum $I_1 \cap I'_2$. Image registration involves four major aspects: similarity metric, transformation, optimization, and interpolation.

1.1.1 Similarity Metric

The similarity metrics include the sum of squared intensity difference, the correlation coefficient, the mutual information (MI), the normalized MI (NMI), and the regional MI (RMI) [6]. MI is the measure of dependence between images and is a powerful metric to determine the similarity between multimodal images. MI is maximum for perfectly matched images. MI M is calculated using (1), where $H(I)$ is the entropy of an image I, and $H(X, Y)$ is the joint entropy of the two images X and Y.

$$M(X, Y) = H(X) + H(Y) - H(X, Y) \tag{1}$$

The NMI N is also a very powerful similarity metric for multimodal image registration. When the images are perfectly matched, the value of $N = 1$. The NMI is determined using (2).

$$N(X, Y) = \frac{H(X) + H(Y)}{H(X, Y)} \tag{2}$$

MI and NMI are the most preferred similarity metrics for intensity-based multimodal image registration, and the other metrics are typically used for monomodal image registration.

1.1.2 Transformation

A transformation function T is used to estimate the geometric relationship between the sensed and the referenced images. The estimate is then used to transform the sensed image. Transformation can be broadly classified as

Rigid transformation: Rigid transformation involves translation and rotation of the image. This is determined using (3) and (4).

$$X = x \cos \theta - y \sin \theta + a \tag{3}$$
$$Y = x \sin \theta + y \cos \theta + b \tag{4}$$

Here, a and b denote translations in x and y dimension, and θ is angle of rotation. X and Y are the transformation parameters.

Non-rigid Transformations: Non-rigid transformations involve translation, rotation, and scaling. Non-rigid transformation is determined using (5) and (6), Here, s denotes the scaling factor.

$$X = xs \cos \theta - ys \sin \theta + a \qquad (5)$$
$$Y = xs \sin \theta + ys \cos \theta + b \qquad (6)$$

1.1.3 Optimization

Medical image registration involves finding the right transformation parameters from a huge set. Since the set is very huge, process of optimization is essential. An optimization algorithm takes a series of intelligent guesses of the transformation parameters, applies them on the sensed image, and uses the similarity metric as the optimization objective function. This metric denotes the degree of accuracy of image registration. Image registration is performed by applying the guessed transformation parameters to the sensed image and determining the objective function on the resultant image and the reference image. The registration process continues by either guessing or obtaining new parameters and recalculating the objective function. This process is repeated until the desired objective function value is reached. An optimization algorithm updates the transformation parameters until the similarity metric between two input images reaches maximum. There are multiple optimization algorithms available in the literature. The traditional optimization algorithms include gradient descent, quasi-Newton, downhill simplex, and simulated annealing and take huge computational time to determine the optimization parameters. In addition, there are many bio-inspired heuristic algorithms, which take comparatively lesser time and less computational resources to determine the optimization parameters, and these include genetic algorithm (GA), particle swarm optimization (PSO) [7], artificial bee colony (ABC) algorithm [8], dragonfly algorithm (DA) [9], and bacterial foraging algorithm (BFA). Many hybrids of these bio-inspired algorithms, such as PSO+GA, PSO+BFA, PSO+neural network (NN), GA+NN, and ABC+NN, have been developed to achieve faster convergence and to minimize the computation time. They are popular in medical image processing as well.

To the best of authors' understanding, there is no comparative performance analysis of modern bio-inspired algorithms applied to non-rigid image registration, and also very few researchers are applying bio-inspired algorithms for medical image registration. Authors aim to bridge this gap by performing a comparative analysis of the performances of three efficient bio-inspired algorithms, namely ABC, DA, and PSO, in non-rigid medical image registration. It is hoped that it benefits researchers and doctors in deciding the best-suiting algorithm according to patients' requirements. The primary contributions of the article are as follows:

- Medical image registration has been recaptured as a continuous optimization problem.
- Bio-inspired algorithms, such as ABC, DA, and PSO, have been used as the tools to approach non-rigid medical image registration.
- Results of ABC-, DA- and PSO-based medical image registration have been presented.
- A comparative investigation of these algorithm has been presented in terms of accurateness and computing time for implementation of image registration.
- A trade-off issue between the quality of registration and computing time has been brought to fore which helps doctors in choosing an appropriate approach.

The remainder of this article has been structured as follows: A survey of literature on medical image registration using bio-inspired techniques has been presented in Sect. 2. Bio-inspired algorithms, such as ABC, DA, and PSO, have been introduced in Sect. 3. Implementation of medical image registration and MATLAB-based numerical simulations using ABC, DA, and PSO have been explained in Sect. 4. Simulation results are presented and discussed in Sect. 5. Finally, conclusions and suggestions for future extension of this research have been presented in Sect. 6.

2 Related Work

Image registration has been tackled using multiple approaches. Common among them is the information theoretic approach MI proposed in [18]. It has been proved as the best deterministic method [6]. When two images are properly aligned, their MI is maximum. The MI approach is robust against noise, sharper peaks at the more correct registration values than other correlation metrics requirement in accurate registration.

Image registration is considered as an ill-posed optimization problem and has been solved using various optimization methods, such as gradient descent, conjugate gradient, quasi-Newton, Gauss-Newton, stochastic gradient descent, Levenberg–Marquardt algorithm, graph-based methods, belief programming, linear programming, and evolutionary methods [10]. Evolutionary computation and other heuristic optimization approaches are more sturdy than the commonly used gradient-based approaches, as they are not dependent on initial solution. In addition, these approaches give specific plans to escape from local minima or maxima. These approaches have been widely used in different kinds of optimization tasks in image registration [11]. Image registration is a high-dimensional problem, computationally very intense, and involves a lot of local minima. Traditional optimization methods are likely to get trapped in local minima. Therefore, researchers have proposed metaheuristic methods to achieve good results [12]. According to a comparative study of evolutionary algorithms for image registration by [10], the best performance has been delivered by a PSO implementation. Multimodal image registration using PSO as optimization technique and NMI as similarity metric has been implemented by [13]. He illustrates the substantial potential of PSO in solving image registration. Also, PSO has proven itself as an efficient optimization algorithm in several areas [14]. The ABC algorithm has been used to solve various constrained, unconstrained, single, and multiple objective optimization problems [8, 15–17]. ABC's performance has been proved to be very efficient in other areas of research [15]. DA is the recent algorithm to join the family of bio-inspired algorithms. It has been used for multilevel segmentation and power system applications and to enhance RFID network lifetime, solar thermal plant efficiency [19–21], etc.

3   Bio-inspired Algorithms for Medical Image Registration

Nature, an affluent source of novel ideas and techniques, inspires scientists and researchers to solve many problems. The fame of nature-inspired algorithm has been attributed to their efficiency, accurate results, simple and humble computation. Three biologically inspired algorithms, namely ABC, DA, and PSO, are explained in the following subsections.

3.1   The ABC Algorithm

The ABC algorithm is an optimization algorithm which draws inspiration from the foraging behavior of natural honeybees [17, 22]. The algorithm involves three different kinds of bees, namely onlooker bees, employed bees, and scout bees. The employed bees perform exploitation of food sources and load the nectar of the source to the hive. The employed bees dance to communicate information about the food source that is being exploited currently. The number of employed bees is equal to the of number food sources. The onlooker bees look at the dance of the employed bees and, based on the dance, find out the amount of nectar in the food source. The scout bees are responsible for the exploration of newer food sources. In the process of exploitation, some food sources might become empty, and the bees which are employed and exploiting these food sources become scouts. The algorithm considers each food source position as a solution to an optimization problem and the nectar amount as the fitness of the solution. The exploration for the food is done by the scout bees and the exploitation by the employed bees. ABC has been applied to solve multiple optimization problems, and a survey on its applications in image, signal, and video processing has been presented in [12, 15].

3.2 The Dragonfly Algorithm

DA draws inspiration from dragonflies [9]. Dragonflies have a unique swarming behavior of swarming only during hunting and migration. Swarming during hunting is called static swarming and that during migration is referred to as dynamic swarming. In case of static swarming, dragonflies create small groups and fly around back and forth over small regions to hunt their preys. Local movements and abrupt changes in the flying path are the main characteristics of a static swarm. In case of dynamic swarming, a huge number of dragonflies make the swarm for migrating in a particular direction over long distances. Static swarm induces exploration behavior, and dynamic swarming induces exploitation behavior. DA is still considered an infant in the bandwagon of swarm algorithms and hence has only few applications in the area of thermal power and photovoltaic systems.

Medical image registration is treated as an optimization problem, and the parameters of optimization are the transformation parameters, namely translation along x and y axes, rotation, and scaling along x and y axes. Thus, the dimensionality of the problem equals five. These parameters are obtained using ABC, DA, and PSO algorithms, and the parameters thus obtained are checked for efficiency using the objective function NMI using (2). Implementations of image registration approached using ABC, DA, and PSO algorithms are described below.

3.3 The PSO Algorithm

PSO is a population-based algorithm [7]. It is inspired by the social behavior of bird flocking and fish schooling. It consists of a swarm of multiple n-dimensional candidate solutions called particles (n is the number of optimal parameters to be determined). Particles explore the search space for a global optimum. Each particle has an initial position, and it moves with some initial velocity. Each particle is evaluated through a fitness function. The fitness value of every particle becomes its personal best, and the minimum (or the maximum) of all particles becomes the global best. All particles try to move toward global best by changing their positions and velocities iteratively. This process is repeated until an acceptable global best is achieved or for a preset number of iterations. The PSO algorithm has been used to solve various optimization problems and has found to be efficient. It has been applied successfully in many engineering domains including wireless sensor networks [14], image registration [23], image segmentation [24], and power systems [25].

4   Numerical Simulation

4.1   Implementation

The ABC-based image registration algorithm, DA-based image registration algorithm, and PSO-based image registration algorithm have been presented in Algorithms 1, 2, and 3, respectively.

4.2   Numerical Simulation

The algorithms presented in this article have been simulated using MATLAB R2012a numeric simulations, on computer having an Intel Core i5 processor. Numerical values pertaining to case studies 1, 2, and 3 are presented in the following subsections:

4.2.1   Case Study 1: ABC-based Medical Image Registration

In this case study, ABC algorithm parameters are set as follows.

- Dimensions $d = 5$
- Abandonment limit $= 0.6 \times D \times P$
- Upper bound of the acceleration coefficient $a = 1$

4.2.2   Case Study 2: DA-based Medical Image Registration

In this case study, DA algorithm parameters are set as follows.

- Separation weight $s = 0.1$, alignment weight $a = 0.1$
- Cohesion weight $c = 0.7$, food factor $f = 1$
- Enemy factor $e = 1$, inertia weight $w = 0.9$

4.2.3   Case Study 3: PSO-based Medical Image Registration

In the study, parameters of PSO are initialized as follows. These are the standard-recommended parameters for the PSO algorithm.

- Dimensions $d = 5$
- Acceleration constants $c_1 = c_2 = 2.0$
- Inertia weight $w = 0.8$

The brain MRI tumorous images considered for simulation are Axial flair, Axial t1 weighted with and without contrast, and Axial t2 weighted and are courtesy of The

Algorithm 1 Pseudocode for the ABC algorithm

1: Initialize food sources m, Dimensions dim, iterations i_{max}, abandonment limit of food source l

2: **for** $i = 1$ to m **do**

3: **for** $j = 1$ to dim **do**

4:   Randomly initialize food source
$$s_i^j = s_{min}^j + r(0, 1) \times (s_{max}^j - s_{min}^j),$$

5: **end for**

6: Trial of each food source $t_i = 0$

7: **end for**

8: $k = 1$

9: **while** $k \leq i_{max}$ **do**

10: Compute $f(s_i)$ using (2) //Employed bees phase of ABC algorithm

11: **for** $i = 1$ to m **do**

12:   **for** $j = 1$ to dim **do**

13:    Exploit novel food source $z_i = s_{ij} + \phi_{ij}(s_{ij} - s_{kj})$

14:   **end for**

15:   Compute novel food fitness value $f(z_i)$ using (2)

16:   **if** $f(z_i) < f(s_i)$ **then**

17:    $s_i = z_i, t_i = 0$ //Make the trial t of the new food source as zero

18:   **else**

19:    $t_i = t_i + 1$

20:   **end if**

21: **end for**

22: Determine the probability for onlooker bees $P_i = \dfrac{F_i}{\sum\limits_{j=1}^{n} F_i}$

23: $t = 0, i = 1$ //Onlooker bees phase of ABC algorithm

24: **repeat**

25:   $r \sim (0, 1)$

26:   **if** $r < P_i$ **then**

27:    $t = t + 1$

28:    **for** $j = 1$ to D **do**

29:     Exploit novel food source $z_i = s_{ij} + \phi_{ij}(s_{ij} - s_{kj})$

30:    **end for**

31:    **if** $f(s_i) > f(z_i)$ **then**

32:     $s_i = z_i, t_i = 0$

33:    **else**

34:     $t_i = t_i + 1$

35:    **end if**

36:   **end if**

37:   $i = i + 1 \ mod(n + 1)$

38: **until** $t = n$

39: **for** $i = 1$ to n **do**

40:   **if** $t_i \geq l$ **then**

41:    $s_i = s_{min}^j + r(0, 1) \times (s_{max}^j - s_{min}^j), t_i = 0$

42:   **end if**

43: **end for**

44: $k = k + 1$

45: **end while**

Algorithm 2 Pseudocode for the DA

1: Initialize alignment weight a, cohesion weight c, food factor f, enemy factor e, inertia weight w Dimensions dim, Separation weight s
2: Randomly Initialize population of n dragonflies D_i
3: Initialize n step vectors δD_i
4: **while do**
5:     Determine the fitness values of all dragonflies
6:     Update enemy and food source
7:     Update w, s, a, c, f, and e
8:     For the current dragonfly D, determine S, A, C, F, and E using

    Separation $Si = -\sigma_{j=1}^{N} D - D_j$

    Alignment $A_i = \frac{\sigma_{j=1}^{N} V_j}{N}$, where V_j is the velocity

    Cohesion $C_i = \frac{\sigma_{j=1}^{N} D_j}{N} - D$

    Food factor $F_i = D^+ - D$, where D^+ is the food source of the current individual
    Enemy factor $E_i = D^- - D$, where D^- is the position of the enemy
9:     // Update neighbouring radius
10:     **if** a dragonfly has at least one neighbouring dragonfly **then**
11:         Update velocity vector using
        $\delta D_{i+1} = (sS_i + aA_i + cC_i + fF_i + eE_i) + w\delta D_i$
12:         Update position vector using
        $D_{i+1} = D_i + \delta D_{i+1}$
13:     **else**
14:         Update position vector using
        $D_{i+1} = D_i + \delta D_{i+1}$
15:     **end if**
16: **end while**

National Library of Medicine, MedPix [26]. These tumorous images of Fig. 2 are input to the bio-inspired algorithms for determining the transformation parameters. The obtained transformation parameters are used to register the images. Results of MATLAB simulation of registering flair image with T1-weighted image using ABC, DA, and PSO are shown in Fig. 3. The results of MATLAB simulation of registering a flair image with a T2-weighted image using ABC, DA, and PSO are shown in Fig. 4. The results of MATLAB simulation of registering a T1-weighted image with a T2-weighted image using ABC, DA, and PSO are shown in Fig. 5. The results of MATLAB simulation of registering a T1 non-contrast image with a T1-weighted image using ABC, DA, and PSO are shown in Fig. 6. Results of MATLAB simulation of registering a T1 non-contrast weighted image with a T2-weighted image using ABC, DA, and PSO are shown in Fig. 7. The results of MATLAB simulation of registering a T1 non-contrast image with a flair image using ABC, DA, and PSO are shown in Fig. 8.

The study has been conducted with various population sizes (20, 30, 50) and iteration value of 500. Thirty trial simulations have been done. Mean of the results of 30 trials has been calculated for inference. Since the algorithms are stochastic, the solutions produced are not the same in all trials, and therefore, the results of multiple trials have been averaged.

Algorithm 3 Pseudocode for the PSO algorithm

1: Initialize maximum allowable iterations r_{max}, target fitness function s to zero and global best value G to maximum value.
2: Set the values of V_{min}, V_{max}, X_{min}, X_{max} where V denotes velocity and X denotes position
3: **for** every particle j **do**
4:     **for** every dimension d **do**
5:         Randomly initialize X_{jd} such that: $X_{min} \leq X_{jd} \leq X_{max}$
6:         Initialize the Personal best values P
7:         $P_{jd} = X_{jd}$
8:         Initialize v_{jd} randomly: $V_{min} \leq v_{jd} \leq V_{max}$
9:     **end for**
10:     Determine $f(X_j)$ using (2)
11:     **if** $f(X_j) < f(G)$ **then**
12:         **for** every dimension d **do**
13:             $G_d = X_{jd}$
14:         **end for**
15:     **end if**
16: **end for**
17: loop $i = 1$
18: **while** $(i \leq r_{max})$ AND $(f(G) > s)$ **do**
19:     **for** every particle j **do**
20:         **for** every dimension d **do**
21:             Calculate velocity $V_{jd}(k)$
22:             $i_1 = c_1 r_{1jd}(k)(Xpbest_{jd}(k) - X_{jd}(k))$
23:             $i_2 = c_2 r_{2jd}(k)(Xgbest_{jd}(k) - X_{jd}(k))$
24:             $V_{jd}(k) = wV_{jd}(k-1) + i_1 + i_2$
25:             Restrict V_{jd} to $V_{min} \leq v_{jd} \leq V_{max}$
26:             Determine position using $X_{jd}(k)$
27:             $X_{jd}(k) = X_{jd}(k) + V_{jd}(k)$
28:             Restrict X_{jd} to $X_{min} \leq X_{jd} \leq X_{max}$
29:         **end for**
30:         Determine $f(X_j)$ using (2)
31:         **if** $f(X_j) < f(G)$ **then**
32:             **for** each dimension d **do**
33:                 $G_d = X_{id}$
34:             **end for**
35:         **end if**
36:     **end for**
37:     $i = i + 1$
38: **end while**

5   Results and Discussion

The brain MRI images (Flair, T1 weighted with contrast, T1 weighted without contrast, and T2 weighted) of same location having tumor shown in Fig. 2 are presented as inputs to the bio-inspired algorithms, and the transformation parameters, namely T_x, T_y, θ, S_x, and S_y, for registration have been determined using ABC, DA, and PSO algorithms. Any two of these images are the inputs to these algorithms (either Flair, T1 weighted with or without contrast, or T2 weighted). The parameters obtained

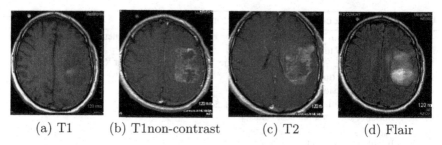

(a) T1   (b) T1non-contrast   (c) T2   (d) Flair

Fig. 2 Brain axial MRI images used for registration

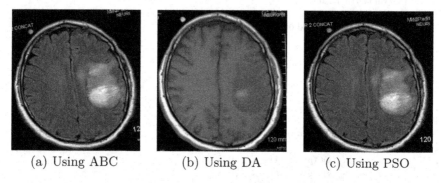

(a) Using ABC   (b) Using DA   (c) Using PSO

Fig. 3 Registration of Flair and T1-weighted images using ABC, DA, and PSO

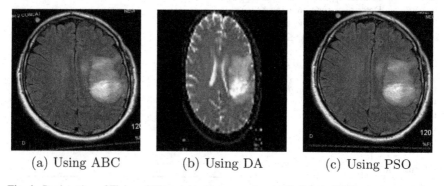

(a) Using ABC   (b) Using DA   (c) Using PSO

Fig. 4 Registration of Flair and T2-weighted images using ABC, DA, and PSO

from the aforementioned algorithms are used for transformations for registration of images. It can be observed from the results presented in figures and tables that the registration using the parameters obtained from the proposed algorithms is very accurate.

The time taken for convergence and the best fitness obtained and the number of iterations taken for convergence for ABC, DA, and PSO algorithms are shown in Table 1. The results of these tables are for 10 trials with a population of 30. It can be

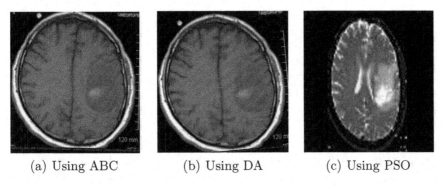

(a) Using ABC   (b) Using DA   (c) Using PSO

Fig. 5 Registration of T2- and T1-weighted images using ABC, PSO, and DA

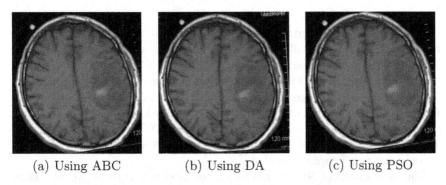

(a) Using ABC   (b) Using DA   (c) Using PSO

Fig. 6 Registration of T2 and T1 non-contrast weighted images using ABC, PSO, and DA

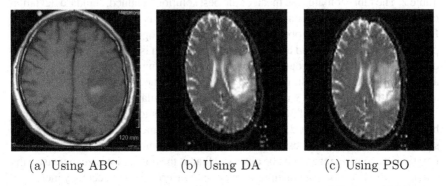

(a) Using ABC   (b) Using DA   (c) Using PSO

Fig. 7 Registration of T2 and T1 non-contrast weighted images using ABC, DA, and PSO

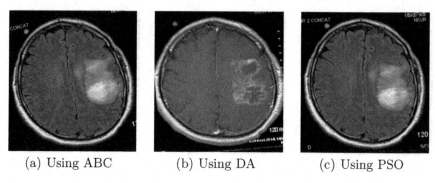

(a) Using ABC   (b) Using DA   (c) Using PSO

Fig. 8   Registration of flair and T1 non-contrast weighted images using ABC, DA, and PSO

observed from Table 1 the fitness value is very close to zero, and also, the time taken for convergence is in terms of seconds, rather than hours which is generally the case with traditional optimization problems.

The plot of best fitness values obtained from these simulations versus iterations for population size of 20 and maximum iteration value of 10000 is shown in Fig. 9 for ABC, DA, and PSO, respectively. It can be observed that the best fitness value reaches very close to zero and also converges after 2000 iterations in all the cases. This denotes the algorithms yield good results for the determination of transformation parameters in image registration.

The mean results of 30 trial runs for varying population size of 20, 30, and 50 and maximum iterations of 500 for the ABC, DA, and PSO algorithms are given in Table 2. The table contains the mean of the best solution obtained, standard deviation of the best solution, mean convergence time, and the mean iterations of convergence. From the table, it can be observed that the obtained optimization transformation parameters are efficient, and also, the standard deviation is very close to zero, which denotes that the obtained optimization parameters using the proposed algorithms are yielding very good solutions close to the mean. The standard deviation of the algorithms becomes better with the increase in the population size.

A statistical summary of 30 trials for different population sizes for the algorithms has been presented in Table 2. These tables contain the mean best solution obtained, standard deviation of the best solution, the mean convergence time, and the mean iterations of convergence. It can be observed that as the population size increases, the efficiency of all the three algorithms in terms of convergence and accuracy increases.

It can be observed from Tables 1 and 2 that DA requires more computing time when compared to ABC and PSO to obtain the optimal results. However, in terms of accuracy of the results, the DA has an upper hand over ABC and PSO algorithms. PSO takes less computing time when compared to the other two, but delivers lowest accuracy of the results. The results point out at a trade-off between accuracy and computing time. Also, as population size increases, the accuracy of all the algorithms improves, but at the cost of increased computing time.

Fig. 9 The plot of the
variation of NMI versus
iterations

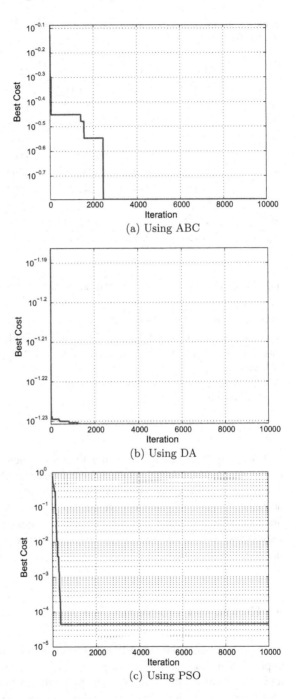

(a) Using ABC

(b) Using DA

(c) Using PSO

Table 1 Simulation results depicting obtained parameters for 10 trial runs and population size of 30 and 500 iterations

Trial	Time taken in seconds			Best fitness		
	ABC	DA	PSO	ABC	DA	PSO
1	1053.4	3671.8	390.66	0.010	6.13E-05	0.019
2	71.70	5124.3	375.25	0.007	3.72E-05	0.558
3	1080.1	3427.5	177.64	0.001	0.0039	0.342
4	650.72	5278.9	413.17	0.004	0.00043	0.505
5	559.33	5312.1	382.89	0.009	0.00045	0.544
6	749.62	2579.8	352.19	0.009	0.00011	0.349
7	469.99	2595.02	341.39	0.008	0.00093	0.562
8	1045.0	3014.5	344.78	0.030	0.00128	0.056
9	474.775	3169.3	298.93	0.008	0.0017	0.008
10	587.4	2782.0	336.47	0.009	0.00067	0.361

Table 2 Results of simulation for 1000 iterations and 30 trials

Algorithm	ABC			PSO			DA		
Population size	20	30	50	20	30	50	20	30	50
Mean fitness	0.3119	0.1404	0.0140	0.425	0.3661	0.104	**0.2147**	**0.210**	**0.1145**
Standard deviation (Fitness)	0.0575	0.0193	0.0016	0.33	0.2436	0.2088	**0.041**	**0.0525**	**0.0217**
Mean computing time (in seconds)	818	1260	2030	**403**	**871**	**1260**	2121	3342	3428

MRI images can be captured using different sequences of T1, T2, and flair. Different sequences of the same pathology give different useful information needed by the doctors, and hence, registration becomes mandatory. Accurate registration results in accurate diagnosis. The results presented in Figs. 3, 4, and 5 show that the images are accurately registered, and the registered images can be used by doctors and the radiologists in analyzing the size of the tumor during its progression and for deciding the course of treatment.

6   Conclusion and Future Work

The research reported in this article has been inspired by the success of bio-inspired, population-based search algorithms to solve the medical image registration problem. The medical image registration problem has been treated as n-dimensional continuous optimization problem. And the optimization parameters are determined using bio-inspired algorithms, namely ABC, DA, and PSO. A brief introduction to

ABC, DA, and PSO algorithm has been provided. A statistical analysis of the results obtained has been presented. The results obtained using the methods are compared.

The results show that ABC, DA, and PSO can handle the image registration successfully. However, there are some trade-off issues. While DA gives higher quality results, it suffers from longer convergence time, whereas PSO results are not as higher quality as ABC and DA, but the convergence is fast. Also, with the increase in population size, better fitness value is obtained but again at the cost of computing time in the cases. The computational complexity of DA is higher than ABC and PSO due to the greater emphasis on exploitation in the search space.

Medical image registration is a one-time exercise. Since medical image registration helps a doctor in decision making, higher quality of results is desirable. Therefore, DA is a better choice than ABC and PSO in situations in which higher precision is needed. However, a dialectical choice between these three algorithms depends on the availability of desired computational speed, computing resources, and accuracy.

This research can be extended further in multiple directions. Variants and hybrids of the bio-inspired algorithms can be developed to achieve improved quality of results and resource efficiency. These algorithms can also be applied to solve three-dimensional medical image registration, deformable image registration, and medical image fusion. Bio-inspired algorithms can also be used to address several unresolved research issues in medical image processing.

Acknowledgements Authors acknowledge with gratitude the support received from REVA University, Bengaluru, and M. S. Ramaiah University of Applied Sciences, Bengaluru. They also express sincere thanks to the anonymous reviewers of this article for their constructive criticism.

References

1. Rueckert, D., Schnabel, J.A.: Registration and segmentation in medical imaging. In: Cipolla R., Battiato, S., Farinella, G. (eds.), Registration and Recognition in Images and Videos, volume 532 of Studies in Computational Intelligence, 137–156. Springer, Heidelberg (2014)
2. Peressutti, D., Gómez, A., Penney, G.P., King, A.P.: Registration of multiview echocardiography sequences using a subspace error metric. IEEE Trans. Biomed. Eng. **64**(2), 352–361 (2017)
3. Xu, R., Athavale, P., Nachman, A., Wright, G.A.: Multiscale registration of real-time and prior MRI data for image-guided cardiac interventions. IEEE Trans. Biomed. Eng. **61**(10), 2621–2632 (2014)
4. Kang, X., Armand, M., Otake, Y., Yau, W.P., Cheung, P.Y., Hu, Y., Taylor, R.H.: Robustness and accuracy of feature-based single image 2-D to 3-D registration without correspondences for image-guided intervention. IEEE Trans. Biomed. Eng. **61**(1), 149–161 (2014)
5. Ebrahimi, M., Kulaseharan, S.: Deformable image registration and intensity correction of cardiac perfusion MRI. In: Proceedings of the 5th International Workshop Statistical Atlases and Computational Models of the Heart-Imaging and Modelling Challenges, Revised Selected Papers, pp. 13–20. Springer, Cham (2015)
6. Tagare, H.D., Rao, M.: Why does mutual-information work for image registration? A deterministic explanation. IEEE Trans. Pattern Anal. Mach. Intell. **37**(6), 1286–1296 (2015)
7. Kennedy, J., Eberhart, R.: Particle swarm optimization. In: Proceedings of the IEEE International Conference on Neural Networks, vol. 4, pp. 1942–1948, Nov 1995

8. Karaboga, D., Akay, B.: A modified artificial bee colony (ABC) algorithm for constrained optimization problems. **11**(3), 3021–3031 (2011)

9. Mirjalili, S.: Dragonfly algorithm: a new meta-heuristic optimization technique for solving single-objective, discrete, and multi-objective problems. Neural Comput. Appl. **27**, 1053–1073 (2015)

10. Sotiras, A., Davatzikos, C., Paragios, N.: Deformable medical image registration: a survey. IEEE Trans. Med. Imaging **32**(7), 1153–1190 (2013)

11. Bermejo, E., Cordón, O., Damas, S., Santamaría, J.: A comparative study on the application of advanced bacterial foraging models to image registration. Inf. Sci. **295**, 160–181 (2015)

12. Damas, S., Cordon, O., Santamaria, J.: Medical image registration using evolutionary computation: an experimental survey. IEEE Comput. Intell. Mag. **6**(4), 26–42 (2011)

13. Schwab, L., Schmitt, M., Wanka, R.: Multimodal medical image registration using particle swarm optimization with influence of the data's initial orientation. In: Proceedings of the IEEE Conference on Computational Intelligence in Bioinformatics and Computational Biology (CIBCB), pp. 1–8, Aug 2015

14. Kulkarni, R.V., Venayagamoorthy, G.K.: Particle swarm optimization in wireless-sensor networks: a brief survey. IEEE Trans. Syst. Man Cybern. Part C Appl. Rev. **41**(2), 262–267 (2011)

15. Akay, B., Karaboga, D.: A survey on the applications of artificial bee colony in signal, image, and video processing. SIViP **9**(4), 967–990 (2015)

16. Brajevic, I.: Crossover-based artificial bee colony algorithm for constrained optimization problems. Neural Comput. Appl. **26**(7), 1587–1601 (2015)

17. Kulkarni, V.R., Desai, V., Kulkarni, R.V.: Multistage localization in wireless sensor networks using artificial bee colony algorithm. In: Proceedings of the IEEE Symposium Series on Computational Intelligence (SSCI), pp. 1–8, Dec 2016

18. Wells, W.M., Viola, P.A., Atsumi, H., Nakajima, S., Kikinis, R.: Multi-modal volume registration by maximization of mutual information. Med. Image Anal. **1**(1), 35–51 (1996)

19. Ganesan, S.I., Manickam, C., Raman, G.R., Raman, G.P.: Dragonfly algorithm based global maximum power point tracker for photovoltaic systems. In: International Conference in Swarm Intelligence, pp. 211–219. Springer, Cham (2016)

20. Murugan, S.: Memory based hybrid dragonfly algorithm for numerical optimization problems. Expert Syst. Appl. **83**, 63–78 (2017)

21. Suresh, V., Sreejith, S.V.: Generation dispatch of combined solar thermal systems using dragonfly algorithm. Computing **99**, 59–80 (2016)

22. Brajevic, I., Tuba, M.: An upgraded artificial bee colony (ABC) algorithm for constrained optimization problems. J. Intell. Manuf. **24**(4), 729–740 (2013)

23. Damas, S., Cordón, O., Santamaria, J.: Medical image registration using evolutionary computation: an experimental survey. IEEE Comput. Intell. Mag. **6**(4), 26–42 (2011)

24. Kulkarni, R.V., Venayagamoorthy, G.K.: Bio-inspired algorithms for autonomous deployment and localization of sensor nodes. IEEE Trans. Syst. Man Cybern Part C Appl. Rev. **40**(6), 663–675 (2010)

25. De Leon-Aldaco, S.E., Calleja, H., Alquicira, J.A.: Metaheuristic optimization methods applied to power converters: a review. IEEE Trans. Power Electron. **30**(12), 6791–6803 (2015)

26. The National Library of Medicine MedPix. https://medpix.nlm.nih.gov/home

Different Length Genetic Algorithm-Based Clustering of Indian Stocks for Portfolio Optimization

Somnath Mukhopadhyay and Tamal Datta Chaudhuri

Abstract In this chapter, we propose a model for portfolio construction using different length genetic algorithm (GA)-based clustering of Indian stocks. First, stocks of different companies, chosen from different industries, are classified based on their returns per unit of risk using an unsupervised method of different length genetic algorithm. Then, the centroids of the algorithm are again classified by the same algorithm. So *vertical* clustering (clustering of stocks by returns per unit of risk for each day) followed by *horizontal* clustering (clustering of the centroids over time) eventually produces a limited number of stocks. The Markowitz model is applied to determine the weights of the stocks in the portfolio. The results are also compared with some well-known existing algorithms. They indicate that the proposed GA-based clustering algorithm outperforms all the other algorithms.

Keywords Different length genetic algorithm · Horizontal clustering
Markowitz model · Portfolio optimization · Return · Risk · Vertical clustering

1 Introduction

A method of selecting an optimal portfolio of stocks was first laid down by Markowitz [22] in which stock returns are quantified as the reward and risk is the variance of the stream of returns over a specific period of time. Konno and Yamazaki [19] used absolute deviation and Speranza [21] used semi-absolute deviation to measure risk in portfolio selection. In these studies of portfolio selection, return and risk are considered as the two fundamental factors that govern investors' choice. In view of this, the multicriteria portfolio selection models [1, 8, 11–14, 20] have generated sufficient interest in researchers.

S. Mukhopadhyay (✉)
Department of Computer Science & Engineering, Assam University, Silchar, India
e-mail: som.cse@live.com

T. D. Chaudhuri
Department of Economics and Finance, Calcutta Business School, Kolkata, India
e-mail: tamalc@calcuttabusinessschool.org

© Springer Nature Singapore Pte Ltd. 2019
J. K. Mandal et al. (eds.), *Advances in Intelligent Computing*,
Studies in Computational Intelligence 687,
https://doi.org/10.1007/978-981-10-8974-9_3

In the recent past, attempts in portfolio formation have either used different algorithms for stock selection or used multi-objective optimization models. Pareto ant colony optimization-based method for multi-objective portfolio selection [5] was proposed in [5]. They proposed an algorithm for finding the approximated solution space. They compared their Pareto ant colony-based meta-heuristics with other heuristic approaches by means of computational experiments with random instances. In [6], portfolio selection was done based on evolutionary algorithm in the Markowitz model. The algorithm finds the portfolio by multi-objective optimization model. The results are compared with other evolutionary multi-objective approaches, and it is shown how these algorithms provide different optimization profiles. The final solution is obtained by Sharpe's index as a measure of risk premium. Anagnostopoulos and Mamanis [2] show that multi-objective evolutionary algorithms are efficient and reliable for portfolio selection and that the performance is not dependent on risk. Barros et al. [4] take a large number of candidate projects as input and formulate a technique which requires analyzing all combinations of candidate projects to find the most effective one.

The literature on portfolio selection has benefited greatly from the use of fuzzy set theory [38, 39] in terms of integrating quantitative and qualitative information, subjective preferences of the investors, and knowledge of the experts. A multi-objective particle swarm optimization algorithm was proposed in [36]. Gupta et al. [15] used an expected value multi-objective model with fuzzy parameters. They considered parameters like short-term return, long-term return, risk, and liquidity as key financial criteria. The solution procedure involved fuzzy goal programming and fuzzy simulation-based real-coded genetic algorithm.

An anticipatory stochastic multi-objective model based on S-Metric maximization was proposed in [3]. For tracking the dynamics of the objective vectors, they proposed an anticipatory learning method so that the estimated S-Metric contributions of each solution can integrate the underlying stochastic uncertainty in the portfolios. An evolutionary multi-objective optimization algorithm for fuzzy portfolio selection was used in [33]. The method optimizes the expected return, the downside-risk, and the skewness of a given portfolio, while taking budget, bound, and cardinality constraints into account. The quantification of the uncertain future return on a given portfolio is approximated by means of LR-fuzzy numbers, while the moments of its return are evaluated using possibility theory.

Clustering is a process of making groups of set of samples or data points so that they become similar within each group. The groups are called clusters [10]. The applications of clustering can be seen in various fields of pattern recognition such as in image processing, object recognition, data mining, machine learning. A popular partitioning clustering algorithm is K-means [35]. The algorithm clusters the samples based on *Euclidean distance* as similarity/dissimilarity measure. The algorithm can cluster large data set, and it is easy to implement. In any fixed-length clustering algorithm like the K-means algorithm, the clustering is obtained by iteratively minimizing a fitness function that is dependent on the distance of the data points to the cluster centers. However, the K-means algorithm, like most of the existing clustering algorithms, assumes a priori knowledge of the number of clusters, K, while in many practical situations, this information cannot be determined in advance. It is also sensitive to the selection of the initial cluster centers and may converge to the local optima. Finding an optimal number of clusters is usually a challenging task, and several researchers have used various combinatorial optimization methods to solve the problem. Some other fixed-length image clustering algorithms [26–28, 37] exist in the literature. Various approaches [17, 18, 23, 32, 34] toward image clustering based on different length chromosome genetic algorithm and different length particle swarm optimization have also been proposed in recent years for clustering. Pakhira et al. [30, 31] proposed some cluster validity indices for fuzzy and crisp clustering.

In this chapter, a real-coded different length genetic algorithm (GA) [9, 16]-based clustering technique is proposed toward selecting a portfolio from 61 Indian stocks over 1445 days. The proposed different length clustering algorithm is first executed vertically, and then horizontally, to determine nine cluster centers, which are actually nine different stocks which form the portfolio. Their weights in the portfolio are optimized using genetic algorithm. In the exercise, risk is minimized subject to total returns greater than 15%, the sum of weights equal to one, and each weight being positive. The rest of the paper is organized as follows. Genetic algorithm is described in Sect. 2. Section 3 describes the proposed different length clustering of stock data. The proposed genetic algorithm-based portfolio selection method and the optimization problem are stated in Sect. 4. The experimental results and discussions are provided in Sect. 5. Section 6 concludes the chapter.

2 Genetic Algorithm

Simple genetic algorithm-based optimization technique works as follows. Initial population of individuals is encoded randomly, and the fitness values of all individuals are evaluated. Until a termination condition is obtained, fittest individuals are selected for reproduction, crossover, and recombination by generating new population in each cascading stage. Extensive simulation is performed using six subprocedures of genetic algorithm. The parameters, attributes, and techniques are discussed in the following Sects. 2.1–2.6.

2.1 Chromosome Encoding and Initial Population

In genetic algorithm-based optimization, the chromosomes are randomly encoded as the candidate solutions.

2.2 Fitness Evaluation

Fitness/objective function is associated with each chromosome. This indicates the degree of goodness of the encoded solution. Fitness of each chromosome is determined by the fitness function. So the system minimizes or maximizes the fitness value in each generation.

2.3 Elitism

In this technique, one copy of the best chromosome is kept inside and outside the population of each generation, which propagates the selection, crossover, and mutation. At the end of each generation, the worst is also selected, and if it is seen that it is better than the best chromosome of the previous generation, then it survives in the population. Otherwise, it is replaced by the best chromosome of the previous generation.

2.4 Selection

In each generation, a mating pool is formed with the same size of the population consisting of better chromosomes through the binary tournament selection (BTS) method [9]. Best chromosome between the two is selected and copied into the mating pool, and the process is repeated until the mating pool is full. Using this method, the chromosome with the lowest fitness value can never be copied into the mating pool. A tie is resolved randomly.

2.5 Crossover

Crossover means exchange of genetic information. It takes place among randomly selected parent chromosomes from the mating pool. Single-point crossover and uniform crossover [24] are the most commonly used schemes. It is a probabilistic operation. It occurs with high probability (μ_c) in each generation. A real value is taken randomly, and if it is seen that it is less than (μ_c), then crossover occurs. Otherwise, the two parents are copied to the next pool for mutation directly. If the operation is performed on two chromosomes, then it also generates two offspring chromosomes. And the process is iterated n/2 times for a mating pool of size n.

2.6 Mutation

It is a process of random alteration in the genetic structure. It introduces genetic diversity into the population by exploring new search areas in the population. Mutating a binary gene involves simple negation of the bit. It is a probabilistic operation. It occurs with very low probability (μ_m) in each generation. A random value is taken first, and if it is seen that it is less than (μ_m), then mutation occurs for the position in the parent chromosome. Otherwise, the mutation does not occur for the particular position.

The parameters which are supplied by user for genetic algorithm are as follows.

- Population size (P)—variable.
- Chromosome length (L)—Fixed/ Variable.
- Probabilities of crossover μ_c, and mutation μ_m. μ_c is kept high and μ_m is kept low.
- Number of generations.

3   Different Length GA (DGA)-Based Clustering

The dimension of the data set used in the paper is 1445×61. It contains the prices of 61 stocks from different sectors for 1445 days. As preprocessing of the data set, 15-day rolling mean of daily returns (Returns) and 15-day rolling standard deviation (Risk) are calculated. The mean is divided by the standard deviation to obtain returns per unit of risk. This data set is then normalized to contain only positive real numbers. These positive real values of dimension 1445×61 are given as parameters to the real-valued different length genetic algorithm for clustering. First, the proposed clustering algorithm is executed for all individual days. This produces cluster centers for all individual days (vertical clustering). Second, the same clustering algorithm is executed on the centers previously obtained for all days taken together (horizontal clustering). This gives ultimately nine cluster centers. These nine centers actually refer to nine different stocks. We then calculated correlation matrix on those nine stocks' returns per unit of risk over the entire time period. After that, an optimization algorithm is executed for finding the weights associated with those nine stocks.

The fitness function proposed by Omran and Salman [26, 28, 29] has been used in the proposed clustering. The proposed fitness function defined in (5) contains three evaluation criteria such as intra-cluster distance measure, inter-cluster distance, and the quantization error minimization function. These criteria are defined respectively in (2), (3), and (4). We consider the same weight of all these three criteria to the fitness of the corresponding chromosomes. The *Euclidean* distance function [28] between i-th data and j-th data in the stocks is computed using (1). It is used to compute intra-cluster distance, inter-cluster distance, and the quantization error. The values of i and j lie in between 1 and 7.

$$d(x) = \sqrt{\Sigma_{i,j}(x_i - x_j)^2} \qquad (1)$$

Let $Z = (z_1, z_2, z_3, \ldots, z_{N_p})$ be the returns of the stocks for n number of individual days. The algorithm maintains a set of chromosomes of population, where each chromosome represents a potential solution to the clustering problem and each chromosome encodes partition of the data set Z. DGA tries to find the number of clusters, N_c. The proposed clustering method has various parameters. N_p, N_c, z_p, m_j, C_j, and $|C_j|$, which are respectively the number of data points to be clustered, number of clusters, p-th data vector, mean or center of cluster j, set of data points in cluster j, and the number of data points in cluster j. Each chromosome can be represented by

$\{m_{i1}, \ldots, m_{ij}, \ldots, m_{iN_c}\}$, where m_{ij} refers to the j-th cluster center vector of the i-th chromosome. In this algorithm, chromosomes have different lengths since the number of clusters is unknown. The chromosomes are initialized with random number of cluster centers in the range $[K_{min}, K_{max}]$, where K_{min} is usually assigned to 2 and K_{max} describes the maximum chromosome length, which represents the maximum possible number of clusters. K_{max} depends on the size and type of data set. DGA, the proposed algorithm, is presented in Algorithm 1.

Algorithm 1 DGA Algorithm

Input: Data: 1445 × 61 Stocks
Output: Partition Matrix
1: **begin**
2:   **for** t=1 to gen **do**   ▷ Total number of generations
3:     **for** I= 1:nch **do**   ▷ Number of Chromosomes
4:       **for** x=1:61 **do**
5:         Let d_i=1;
6:         **for** l_p=1:lch(I) **do**   ▷ Length of ith chromosome
7:           v=ch(I,lp);
8:           val = data(v,:);
9:           dd=data(x);
10:          d(di)= edist(dd,val);   ▷ Euclidean distance using 1
11:          d_i=d_i+1;
12:         **end for**
13:         m_i=mini(d,lch(I));
14:         Save m_i in pmat;
15:       **end for**
16:       intra(I)=intradist(data,pmat,lch(I),ch,I);   ▷ using 2)
17:       qe(I)=qerror(data,cmat,lch(I),ch,I);   ▷ using (4)
18:       inter(I)=interdist(ch,lch(I),data,I);   ▷ using (3)
19:       f(I)=intra(I) 1-inter(I) + qe(I);   ▷ using (5)
20:     **end for**
21:     **if** t==1 **then**   ▷ Elitism Model
22:       bes=Best Chromosome number of Current Gen
23:     **end if**
24:     **if** t≥2 **then**
25:       wor=Worst Chromosome number of Current Gen
26:       bes=f(bestchno);   ▷ Best Chromosome of Current generation
27:       **if** bes<wor **then**   ▷ If Prev Best is better than Current Worst
28:         ch(wchno,:)=bes;   ▷ Then Prev Best replaces the Current Worst
29:       **end if**
30:     **end if**
31:     Selection using Binary Tournament Selection (BTS)
32:     Crossover on Real Coded Different length Chromosome (Algorithm 2)
33:     Real Coded Mutation (Algorithm 3)
34:     Boundry Restriction
35: **end for**
36: **end**

The intra-cluster distances of all the clusters are measured, and the maximum one among all the clusters is selected in d_{max} which is defined in (2), where Z is a partition matrix representing the assignment of data points to clusters of chromosome i. A smaller value of d_{max} means that the clusters are more compact.

$$d_{max}(Z, x_i) = \max_{j=1\ to\ N_c} \{ \sum_{\forall z_p \in C_{ij}} d(z_p, m_{ij}) / |C_{ij}| \} \qquad (2)$$

Inter-cluster separation distances for all clusters are measured, and the minimum distance between any two clusters is calculated using (3). A large value of d_{min} means that the clusters are well separated.

$$d_{min}(x_i) = \min_{\forall j1,j2,j1 \neq j2} \{ d(m_{ij_1}, m_{ij_2}) \} \qquad (3)$$

The quantization error function [7, 28] is proposed in the clustering of stock market data points, which calculates the average distance of the data points of a cluster to its cluster centers, followed by the average distances of all clusters. Then, a new average is calculated. The problem of Esmin et al. [7] is that any cluster with one data point would affect the final result with another cluster containing many data points. Suppose, for i-th chromosome in a cluster which has only one data point and very close to the center, there is another cluster that has many data points which are not so close to the center. The problem has been resolved by assigning less weight to the cluster containing only one data point than with cluster having many data points. The weighted quantization error function is given in (4), where N_0 is the total number of data vectors to be clustered. The fitness function is constructed by intra-cluster distance d_{max}, inter-cluster distance d_{min} along with the quantization error Q_e function. The fitness function is used to minimize $f(x_i, Z)$ [37] which is given in (5). Here z_{max} is a big value assumed. In the optimization function, equal weights are assigned to the three distance functions. The fitness function is given to the optimization technique and which minimizes the value of f in each generation to make the stock market data well clustered.

$$Q_e = \{ \sum_{\forall j=1\ to\ N_c} [(\sum_{\forall z_p \in c_{ij}} d(z_p, m_{ij}) / |C_{ij}| \cdot (N_0 / |C_{ij}|)] \} \qquad (4)$$

$$f(x_i, Z) = d_{max}(z, x_i) + (z_{max} - d_{min}(x_i)) + Q_e \qquad (5)$$

The proposed crossover technique on real-coded and different length chromosomes is given in Algorithm 2. As the operation is on different length chromosomes, it either discards or adds some positional values to make them same length. The operation starts with checking with a random number with the crossover probability (μ_c) to go for the crossover operation. After that again a random number is taken for making the crossover operation. Depending on the value of the random number, some positional values are either discarded and zero appended. When they are of same length, the mean is calculated for generating the offspring.

The proposed mutation operation on real-coded chromosome is given in Algorithm 3. The operation is performed on each chromosome with low probability μ_m. A marginal change in the chromosomes in the mutation operation is based on a random string, *velocity*, represented by v. The maximum value v_{max} of v is empirically chosen as 0.25, and it is linearly decreased in each generation to v_{min}, 0.01. The input chromosomes, output offspring, velocity v, and probabilities μ_m are all taken in the form of strings. The mutation operation is done in each position based on the probability for each position. A random number is generated for each position and compared to the mutation probability to go through the mutation operation. The proposed operation adds the random velocity to the parent chromosome to make the offspring.

Algorithm 2 Crossover technique on Real Coded Different length Chromosomes

---

Input: Chromosomes p_1 and p_2
Output: Offsprings os_1 and os_2
1: **begin**
2: Let r=rand()   ▷ Random number in between 0 and 1
3: **if** $|(p_2)| > |(p_1)|$ **then**   ▷ Length function
4:      swap(p_1,p_2);   ▷ Keeping the larger chromosome in p_1 always
5: **end if**
6: $l_1=|p_1|$;
7: $l_2=|p_2|$;
8: Let d=l_1-l_2;
9: $x_1=p_1$*r;
10: $y_1=p_2$*(1-r);
11: $x_2=p_1$*(1-r);
12: $y_2=p_2$*r;
13: **if** r $\geq$ 0.5 **then**
14:      $y_1=[y_1$,zeros(1,d)];   ▷ d number of zero appending to make them similar length
15:      os_1=mean(x_1+y_1);
16:      $x_2=x_2(1:l_2)$;   ▷ deletion from end to make them similar length
17:      os_2=mean(x_2+y_2);
18: **else**
19:      $x_1=x_1(1:l_2)$;
20:      os_1=mean(x_1+y_1);
21:      $y_2=[y_2$,zeros(1,d)];
22:      os_2=mean(x_2+y_2);
23: **end if**
24: **end**

Algorithm 3 Mutation technique on Real Coded Chromosome

Input: Chromosome p as String
Output: Offspring os as String
 1: **begin**
 2: Let μ_m be the string of mutation probabilities.
 3: Let v be the velocities which make changes in the chromosomes in each generation. It is randomly chosen in $[0, v_{max}/v_{min}]$. v_{max} and v_{min} are empirically set in between 0.25 to 0.01. The value of v_{max} is decreased linearly in each generation to eventually v_{min}.
 4: Offspring os is calculated using (4).

$$os = p + \mu_m * v \qquad (6)$$

 5: **end**

---

4 Portfolio Selection and Optimization

The objective function to be minimized by genetic algorithm is given in (7). The detailed algorithm is given in Algorithm 4;

$$\min : W^T \times \Sigma \times W \qquad (7)$$

subject to contraints:
$W^T \times R = \theta,$
$\Sigma W_i = 1, \forall i$
and
$W_i > 0, \forall i.$ where

- θ is the required return, taken as 0.15/250.
- Σ is risk covariance matrix (n × n).
- R (n × 1) returns of stocks.
- W^T (n × 1) weights of stocks.

Algorithm 4 Portfolio Optimization Technique

Input: Chromosome p as String
Output: Offspring os as String
 1: **begin**
 2: Encode Chromosome p:
 3: $p_j = w_i = r_i$, where $r_i \in [0,1]$; $\Sigma r_i = 1$; j is number of chromosomes; $1 \le i \le 9$;
 4: Fitness Calculation :
 5: $f_i - W^T * \Sigma * W$; where $W^T * R - 0.0006$
 6: Binary Tournament Selection
 7: Probabilistic Uniform Crossover
 8: Probabilistic Mutation Operation
 9: **end**

---

5   Results and Comparisons

The performance of the proposed algorithm is measured by three evaluation metrics, *intra-cluster distance, inter-cluster distance*, and *quantization error*. The performance of the proposed clustering algorithm is compared with five existing algorithms, namely *K-means*, Man et al. [37], *FPSO* [28], *DPSO* [25], and *VPSO* [26]. For comparison purpose, the following parameter values are used:

- Number of particles (NOP)/chromosomes = 20
- Maximum number of clusters = 50
- maximum number of iterations = 200.

Number of iterations for Man et al., FPSO, DPSO, VPSO, and the proposed hybrid model are set to 200. For K-means, the number of iterations will be 200 × number of particles, because in each iteration the fitness of 20 particles are computed in clustering algorithms. For FPSO, DPSO, VPSO, and proposed algorithm, minimum (K_{min}) and maximum (K_{max}) number of clusters are set to 2 and 50, respectively. For K-means algorithm, only mean square error (MSE) is used as fitness function. For Man et al., FPSO, DPSO, VPSO, and proposed algorithm, the same fitness function is used for evaluation which is given in (5).

Table 1 shows the intra-cluster distance d_{max}, inter-cluster distance d_{min}, weighted quantization error Q_e, and fitness value by the existing and proposed algorithms. The results in this table are based on the vertical clustering on the data set mentioned. The intra-cluster distance and quantization error values are minimized in each generation by the proposed algorithm, whereas the inter-cluster distance is maximized. The fitness value is also minimized as a whole. For K-means algorithm, from the results we can see that all the three performance metrics like intra-cluster distance, inter-cluster distance, and weighted quantization error are very poor compared to all other algorithms. Man et al., FPSO, and DPSO perform much better than the K-means algorithm in terms of all performance metrics. The proposed hybrid model of PSO and GA is best with respect of all those metrics. It obtains a significant less quantization error value compared to all other algorithms in the table. The last column of the table shows the fitness values of all the algorithms. We can see from

Table 1 Clustering results using intra distance (d_{max}), quantization error (Q_e), inter distance (d_{min}), and fitness value

Algorithm	d_{max}	Q_e	d_{min}	Fitness value
(1) K-means	0.9678	0.4531	0.0471	1.4680
(2) Man et al.	0.8793	0.1202	0.1892	1.1887
(3) FPSO	0.8102	0.0955	0.2167	1.1224
(4) DPSO	0.7201	0.0676	0.2822	1.0699
(5) VPSO	0.4954	0.0207	0.5043	1.0204
(6) Proposed	0.1623	0.0031	0.7342	0.8996

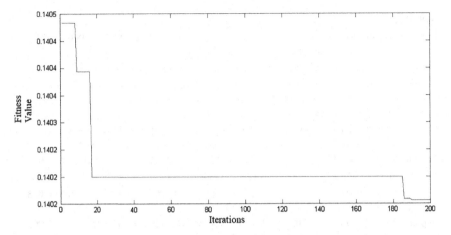

Fig. 1 Bar diagram showing minimization of fitness values in 200 iterations

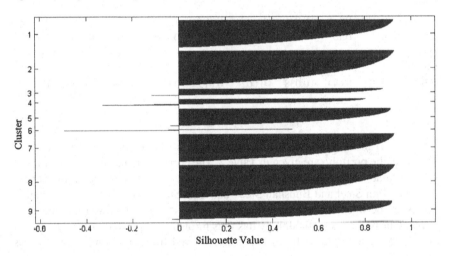

Fig. 2 Silhouette plot of the data points

the table that the proposed GA-based clustering algorithm outperforms all the other algorithms.

The proposed algorithm for clustering of stocks has been executed for 200 number of generations. In Fig. 1, we have shown the *gBest* values when the algorithm iterates for 200 number of generations. We can see from these figures that the proposed algorithm minimizes the global best fitness values in successive iterations.

In Fig. 2, we have given the cluster silhouettes which validate the proposed clustering algorithm. In obtaining the silhouette values, Euclidean distance measure is used. From the figure, we can see that very few points have a low or negative silhouette values, where most of the points have positive and high silhouette values.

Table 2 Weight vector

Stock	Emami	BHEL	JK Paper	Supreme	TATA Steel	Bhushan	WIPRO	Godrej Con	IFB Ind
Weight	0.15	0.05	0.06	0.15	0.05	0.09	0.15	0.15	0.15

Table 3 Correlation matrix

Correl	Emami	BHEL	JK Paper	Supreme	TATA Steel	Bhushan	WIPRO	Godrej Con	IFB Ind
Emami	1	0.057	0.115	0.073	0.024	−0.029	0.015	0.027	−0.021
BHEL	0.057	1	0.097	0.122	0.396	0.176	0.135	0.136	0.044
J K Paper	0.115	0.097	1	0.026	0.057	0.075	−0.020	−0.002	0.038
Supreme	0.073	0.122	0.026	1	0.173	0.132	0.102	0.123	0.002
TATA Steel	0.024	0.396	0.057	0.173	1	0.307	0.207	0.164	0.015
Bhushan	−0.029	0.176	0.075	0.132	0.307	1	0.140	0.105	0.035
WIPRO	0.015	0.135	−0.020	0.102	0.207	0.140	1	0.087	0.003
Godrej Con	0.027	0.136	−0.002	0.123	0.164	0.105	0.087	1	−0.046
IFB Ind	−0.021	0.044	0.038	0.002	0.015	0.035	0.003	−0.046	1

Out of 61 stocks, the name of the 9 selected stocks for the portfolio and their weights in the portfolio are given in Table 2. The correlation matrix given in Table 3 indicates that the pairwise correlation of the returns is quite low, and at times negative. For Tata Steel and Bhushan Steel, and Tata Steel and BHEL, the correlation figures are between 0.3 and 0.4. Otherwise, the rest of the figures are almost negligible. Our proposed methodology has thus thrown up a portfolio of stocks which are not correlated with each other and has evolved from an optimization process involving both vertical clustering and horizontal clustering.

6 Conclusion

This paper proposes a real-coded different length genetic algorithm-based clustering model for selecting a portfolio of Indian stocks. An optimization exercise was performed for selecting the weights of each stock in the portfolio. The data set taken is for 1445 days for 61 stocks. Vertical clustering followed by horizontal clustering and then selecting the portfolio is done by the proposed methodology. The proposed clustering algorithm is validated by comparing it with other state-of-the-art algorithms in the literature. The silhouette plot shows that the proposed algorithm obtains excellent silhouette values. The correlation matrix shows that the algorithm chose stocks that have low correlation.

References

1. Abdelaziz, F.B., Masmoudi, M.: A multiple objective stochastic portfolio selection problem with random beta. Int. Trans. Oper. Res. **21**(6), 919–933 (2014). https://doi.org/10.1111/itor.12037
2. Anagnostopoulos, K.P., Mamanis, G.: Multiobjective evolutionary algorithms for complex portfolio optimization problems. Comput. Manag. Sci. **8**(3), 259–279 (2011). https://doi.org/10.1007/s10287-009-0113-8
3. Azevedo, C.R.B., Zuben F.J.V.: Anticipatory stochastic multi-objective optimization for uncertainty handling in portfolio selection. In: 2013 IEEE Congress on Evolutionary Computation, pp. 157–164 (2013). https://doi.org/10.1109/CEC.2013.6557566
4. Barros, M.D.O., Costa, H.R., Figueiredo, F.V., Rocha, A.R.C.D.: Multiobjective optimization for project portfolio selection. In: Proceedings of the 14th Annual Conference Companion on Genetic and Evolutionary Computation GECCO '12, pp. 1541–1542 (2012). ACM, New York, NY, USA. https://doi.org/10.1145/2330784.2331037
5. Doerner, K., Gutjahr, W.J., Hartl, R.F., Strauss, C., Stummer, C.: Pareto ant colony optimization: a metaheuristic approach to multiobjective portfolio selection. Ann. Oper. Res. **131**(1), 79–99 (2004). https://doi.org/10.1023/B:ANOR.0000039513.99038.c6
6. Duran, C.F., Cotta, C., Fernández, A.J.: Evolutionary optimization for multiobjective portfolio selection under Markowitz's model with application to the caracas stock exchange, pp. 489–509. Springer, Berlin, Heidelberg (2009). https://doi.org/10.1007/978-3-642-00267-0_18
7. Esmin, A.A.A., Pereira, D.L., de Arajo, F.P.A.: Study of different approach to clustering data by using particle swarm optimization algorithm. In: Proceedings of the IEEE World Congress on Evolutionary Computation (CEC, 2008), pp. 1817–1822. Hong Kong, China (2008)
8. Fang, Y., Lai, K., Wang, S.Y.: Portfolio rebalancing model with transaction costs based on fuzzy decision theory. Eur. J. Oper. Res. **175**(2), 879–893 (2006). https://doi.org/10.1016/j.ejor.2005.05.020, http://www.sciencedirect.com/science/article/pii/S0377221705005102
9. Goldberg, G.: Genetic algorithm in search, optimization and machine learning. Addison-Wesley (1989)
10. Gose, E., Johnsonbough, R., Jost, S.: Pattern recognition and image analysis. Prentice-Hall (1996)
11. Gupta, P., Inuiguchi, M., Mehlawat, M.K.: A hybrid approach for constructing suitable and optimal portfolios. Exp. Syst. Appl. **38**(5), 5620–5632 (2011). https://doi.org/10.1016/j.eswa.2010.073, http://www.sciencedirect.com/science/article/pii/S0957417410012170
12. Gupta, P., Mehlawat, M.K., Mittal, G.: Asset portfolio optimization using support vector machines and real-coded genetic algorithm. J. Global Opt. **53**(2), 297–315 (2012). https://doi.org/10.1007/s10898-011-9692-3
13. Gupta, P., Mehlawat, M.K., Saxena, A.: A hybrid approach to asset allocation with simultaneous consideration of suitability and optimality. Inf. Sci. **180**(11), 2264–2285 (2010). https://doi.org/10.1016/j.ins.2010.02.007, http://www.sciencedirect.com/science/article/pii/S0020025510000630
14. Gupta, P., Mehlawat, M.K., Saxena, A.: Asset portfolio optimization using fuzzy mathematical programming. Inf. Sci. **178**(6), 1734–1755 (2008). https://doi.org/10.1016/j.ins.2007.10.025, http://www.sciencedirect.com/science/article/pii/S0020025507005221
15. Gupta, P., Mittal, G., Mehlawat, M.K.: Multiobjective expected value model for portfolio selection in fuzzy environment. Opt. Lett. **7**(8), 1765–1791 (2013). https://doi.org/10.1007/s11590-012-0521-5

16. Holland, J.H.: Adaptation in Natural and Artificial Systems. MIT Press, Cambridge, MA, USA (1992)
17. Javier, L., Laura, L., Armando, G.: Varmopso: multi-objective particle swarm optimization with variable population size. In: Advances in Artificial Intelligence IBERAMIA. Lecture Notes in Computer Science, vol. 6433. Springer, Berlin, Heidelberg (2010)
18. Katari, V., Ch, S., Satapathy, R., Ieee, M., Murthy, J., Reddy, P.P.: Hybridized improved genetic algorithm with variable length chromosome for image clustering abstract. Int. J. Comput. Sci. Netw. Secur. **7**(11), 121–131 (2007)
19. Konno, H., Yamazaki, H.: Mean-absolute deviation portfolio optimization model and its applications to Tokyo stock market. Manage. Sci. **37**(5), 519–531 (1991). https://doi.org/10.1287/mnsc.37.5.519
20. Li, X., Qin, Z., Kar, S.: Mean-variance-skewness model for portfolio selection with fuzzy returns. Eur. J. Oper. Res. **202**(1), 239–247 (2010). https://doi.org/10.1016/j.ejor.2009.05.003, http://www.sciencedirect.com/science/article/pii/S0377221709003154
21. Mansini, R., Ogryczak, W., Speranza, M.G.: Linear models for portfolio optimization, Springer International Publishing, Cham, pp. 19–45. https://doi.org/10.1007/978-3-319-18482-1_2
22. Markowitz, H.: Portfolio selection. J. Financ. **7**(1), 77–91 (1952). https://doi.org/10.1111/j.1540-6261.1952.tb01525.x
23. Maulik, U., Bandyopadhyay, S.: Fuzzy partitioning using a real-coded variable-length genetic algorithm for pixel classification. IEEE Trans. Geosci. Remote. Sens. **41**(5), 1075–1081 (2003). https://doi.org/10.1109/TGRS.2003.810924
24. Michalewicz, Z.: Genetic algorithms + data structures = evolution programs, Springer (1996)
25. Mukhopadhyay, S., Mandal, P., Pal, T., Mandal, J.K.: Image Clustering Based on Different Length Particle Swarm Optimization (DPSO), Springer International Publishing, Cham, pp. 711–718 (2015). https://doi.org/10.1007/978-3-319-11933-5_80
26. Mukhopadhyay, S., Mandal, J.K., Pal, T.: Variable length PSO-based image clustering for image denoising. In: Handbook of Research on Natural Computing for Optimization Problems. IGI Global, Hershey (2016)
27. Mukhopadhyay, S., Mandal, J.K.: Adaptive median filtering based on unsupervised classification of pixels. In: Handbook of Research on Computational Intelligence for Engineering, Science and Business. IGI Global, Hershey (2013)
28. Mukhopadhyay, S., Mandal, J.K.: Denoising of digital images through pso based pixel classification. Central Eur. J. Comput. Sci. **3**(4), 158–172 (2013)
29. Omran, M., Engelbrecht, A.P., Salman, A.: Particle swarm optimization method for image clustering. Int. J. Pattern Recognit. Artif. Int. **19**, 297–322 (2005)
30. Pakhira, M.K., Bandyopadhyay, S., Maulik, U.: A study of some fuzzy cluster validity indices, genetic clustering and application to pixel classification. Fuzzy Sets Syst. **155**(2), 191–214 (2005). https://doi.org/10.1016/j.fss.2005.04.009, http://www.sciencedirect.com/science/article/pii/S0165011405001661
31. Pakhira, M.K., Bandyopadhyay, S., Maulik, U.: Validity index for crisp and fuzzy clusters. Pattern Recognit. **37**(3), 487–501 (2004). https://doi.org/10.1016/j.patcog.2003.06.005, http://www.sciencedirect.com/science/article/pii/S0031320303002838
32. Qiu, M., Liu, L., Ding, H., Dong, J., Wang, W.: A new hybrid variable-length GA and PSO algorithm in continuous facility location problem with capacity and service level constraints. In: IEEE/INFORMS International Conference on Service Operations, Logistics and Informatics, 2009. SOLI '09, pp. 546–551 (2009)
33. Saborido, R., Ruiz, A.B., Bermdez, J.D., Vercher, E., Luque, M.: Evolutionary multi-objective optimization algorithms for fuzzy portfolio selection. Appl. Soft Comput. **39**, 48–63. https://doi.org/10.1016/j.asoc.2015.11.005, http://www.sciencedirect.com/science/article/pii/S1568494615007164
34. Srikanth, R., George, R., Warsi, N., Prabhu, D., Petry, F., Buckles, B.: A variable-length genetic algorithm for clustering and classification. Pattern Recognit. Lett. **16**(8), 789–800 (1995). https://doi.org/10.1016/0167-8655(95)00043-G, http://www.sciencedirect.com/science/article/pii/016786559500043G

35. Tan, P., Steinbach, M., Kumar, V.: Introduction to data mining, Pearson Education (2006)
36. Wang, B., Watada, J.: Multiobjective particle swarm optimization for a novel fuzzy portfolio selection problem. IEEJ Trans. Electr. Electron. Eng. **8**(2), 146–154 (2013). https://doi.org/10.1002/tee.21834
37. Wong, M.T., He, X., Yeh, W.C.: Image clustering using particle swarm optimization. In: 2011 IEEE Congress on Evolutionary Computation (CEC), pp. 262–268 (2011)
38. Zadeh, L.: Fuzzy sets. Inf. Control **8**(3), 338–353 (1965). https://doi.org/10.1016/S0019-9958(65)90241-X, http://www.sciencedirect.com/science/article/pii/S001999586590241X
39. Zadeh, L.A.: Toward a generalized theory of uncertainty (GTU) an outline. Inf. Sci. **172**(12), 1–40 (2005). https://doi.org/10.1016/j.ins.2005.01.017, http://www.sciencedirect.com/science/article/pii/S002002550500054X

An Evolutionary Matrix Factorization Approach for Missing Value Prediction

Sujoy Chatterjee and Anirban Mukhopadhyay

Abstract Sparseness of data is a common problem in many fields such as data mining and pattern recognition. During the last decade, collecting opinions from people has been established to be an useful tool for solving different real-life problems. In crowdsourcing systems, prediction based on very few observations leads to complete disregard for the inherent latent features of the crowd workers corresponding to the items. Similarly in bioinformatics, sparsity has a major negative impact in finding relevant gene from gene expression data. Although this problem is being studied over the last decade, there are some benefits and pitfalls of the different proposed approaches. In this article, we have proposed a genetic algorithm-based matrix factorization technique to estimate the missing entries in the rating matrix of recommender systems. We have created four synthetic datasets and used two real-life gene expression datasets to show the efficacy of the proposed method in comparison with the other state-of-the-art approaches.

Keywords Matrix factorization · Sparsity · Genetic algorithm
Judgment analysis

1 Introduction

Over the years, seeking user opinions in collaborative filtering system to recommend similar types of users has been proved to very effective [1, 6]. Recently, it is seen that engaging multiple non-experts can produce better results rather than involving few experts [13] to solve complex real-life problems. This betterment of result can be in terms of quality, financial as well as temporal cost. In crowdsourcing environment, a

S. Chatterjee (✉) · A. Mukhopadhyay
Department of Computer Science & Engineering, University of Kalyani,
Nadia 741235, India
e-mail: sujoy@klyuniv.ac.in

A. Mukhopadhyay
e-mail: anirban@klyuniv.ac.in

© Springer Nature Singapore Pte Ltd. 2019
J. K. Mandal et al. (eds.), *Advances in Intelligent Computing*,
Studies in Computational Intelligence 687,
https://doi.org/10.1007/978-981-10-8974-9_4

requester who needs to solve a particular task can post the job in an open platform to seek crowd opinions. Now while attempting the question, crowd workers can select any number of questions in which they feel that they are capable of answering those properly. Thus, their opinions are stored in a response matrix where the rows represent the crowd workers and the columns represent the questions. The cell value of the response matrix contains the option of a particular crowd worker for a particular question. As there is no mandatory rule for attempting the questions, the response matrix in general becomes very much sparse. Due to this sparsity problem in crowdsourcing domain, aggregation from multiple opinions to find a robust consensus can be heavily affected.

In crowdsourcing, involving non-experts instead of experts can produce better result while making the final judgment [3, 8, 13, 25]. In this environment, at some situations, the number of opinions collected from crowd for a question is very low. Thus, it creates a problem to find the accuracy of the crowd workers due to such small set of attempted questions. Eventually, the calculated accuracy of the crowd workers depending only on a few set of questions may not be appropriate. Moreover, assigning a task to the crowd workers cannot be accomplished correctly if the requester relies on those workers attempting very few questions. Alternatively if the sparseness of the response matrix can be reduced wisely, then proper assignment of task as well as task-routing can be solved effectively.

In bioinformatics, microarray technology is one of the widely used tools for monitoring gene expression levels of an organism. It is mainly used for cancer classification, identification of relevant gene for diagnosis, and studying drug effect. However, there are still challenges in it due to the presence of uncharacterized variables that need proper interpretation of data. Gene expression data usually suffer from this problem of missing values due to several reasons such as low resolution, image noise, hybridization failures, etc. Besides it, the wrong prediction of missing value leads to loss of informative genes. Therefore, to properly identify the relevant genes as well as to enhance the quality of overall prediction, sophisticated models are needed to be developed. Nonnegative matrix factorization (NMF) [10, 12, 18] can be considered as a fundamental tool to alleviate this problem. During the last decade, a significant amount of research is being carried out on nonnegative matrix factorization [11]. Although there are some improvements in state-of-the-art approaches, there exist some inherent problems due to the random initialization of the matrices. The final matrix might be highly dependent on the initial matrices. Again, if the methods can be repeatedly executed with a goal to extract the best result, then it can cost a large amount of time. These problems motivate us to propose a new method in the framework of genetic algorithm for lifting up the performance of nonnegative matrix factorization.

In this paper, the missing value prediction is framed in an optimization model and a genetic algorithm-based method [21] is provided for mitigating the problem of filling up the missing values. To speed up the search process, we have used a local refinement step by using gradient descent method after the creation of initial population. Here, root mean squared error (RMSE) is used as the objective function to find the closeness of the nonzero elements of the original matrix and the predicted matrix obtained after application of this method. The performance of the proposed method is demonstrated by applying it on four artificial datasets as well as two real-life gene expression datasets. The results show the effectiveness of it over the state-of-the-art approaches. A preliminary version of this work appears in [5]. In this current work, we address this problem applying on more datasets with larger size with minor changes in local refinement step.

2 Matrix Factorization for Missing Value Prediction

The objective of matrix factorization is to remove the error due to the large scale of sparsity. We start with an example matrix normally used in recommender system. This model tries to map the users and the items into a latent space so that the user-item is modeled as an inner product.

Suppose there is a set of users U and a set of items I. Let again R be the rating matrix that contains users' interests over the items and the dimension of R is $|U \times I|$. Now to capture the latent feature, the objective of the matrix factorization method is to find two matrices for an integer $K \leq \min\{|U|, |I|\}$ $W \in \mathbb{R}^{|U \times K|}$, $H \in \mathbb{R}^{|I \times K|}$, such that the product of W and H^T is approximately equal to R. Mathematically, it is denoted by

$$R \approx W \times H^T = \hat{R}. \tag{1}$$

Now to measure the goodness of the solution, the common measures used are Frobenius norm and Kullback–Leibler divergence [15]. To illustrate this, suppose for a nonnegative matrix V, the objective of NMF is to compute nonnegative matrix factors W and H so that $V \approx WH$. The aim is to minimize $J(W, H)$ as defined in the following equation.

$$\min_{W \geq 0, H \geq 0} J(W, H) = \frac{1}{2} ||V - WH||_F^2. \tag{2}$$

Here, $||.||_F$ is the Frobenius norm. In the above equation, both the matrices should have nonzero elements. Hence, the product of W and H approximates the unknown entries of original matrix V based on the known entries, and thus, missing values can be predicted efficiently.

3 Related Work

A significant volume of research has been accomplished over the years to study different issues dealing with nonnegative matrix factorization problem. There are so many applications like data mining, bioinformatics, information retrieval, where NMF can be used as an effective technique [4, 26]. The first NMF method was introduced by Paatero and Teppar [22, 23], but the limitation of their method is that they did not consider the negativity constraint. They focussed on positive matrix factorization (PMF) model and investigated on few important aspects of PMF in order to make robust factorization using Huber influence function. After that the researchers focussed on this problem from different angles. Among them, multiplicative update algorithms [15] is one the popular algorithms and it provides better solutions with strong convergence proof. Here, multiplicative update is modified by rescaled gradient descent.

Another method, namely hierarchical alternating least squares (HALS) algorithm, has been introduced with convergence proof. Several methods motivated by alternating nonnegative least squares (ANLS) framework [16] have been introduced for matrix factorization problem and for solving Eq. 2. In this framework, the variables are kept into two subgroups and those groups are updated iteratively. The steps of this method are described below.

- Fill up the matrix $H \in \mathbb{R}$ with nonnegative elements.
- The following equation is solved iteratively until some convergence criteria are met.

  - $\min\limits_{W>0} ||H^T W^T - A^T||_F^2$, when H is kept fixed and
  - $\min\limits_{H>0} ||WH - A||$, when W is fixed.
  - Normalize the columns of W, and the rows of H are scaled accordingly.

In this method, to solve the original problem, primarily two problems are solved. Active set method is the solution to solve this. Here, exchanges of variables between two working sets happen, and the two-block coordinate descent algorithm is used. The limitation of the method is that it executes at slower speed when the number of unknown parameters gets high. However, it provides the convergence proof. Different techniques have also been proposed to speed up the convergence rate of the algorithms. Different techniques such as conjugate gradient and projected gradient are also popular. The NMF algorithms can be categorized into four subclasses, such as (i) semi-NMF, (ii) nonnegative tensor factorization (NTF), (iii) nonnegative matrix-set factorization (NMSF), and (iv) kernel NMF (KNMF). In general, NMF restricts every element of the data matrix R should be nonnegative. But the semi-NMF method proposed in [9] allows no restriction on the signs of one matrix H of the two decomposed matrices W and H. Another extension of the NMF methods has been proposed in [27] to consider the original data matrix as factorized form of three matrices. In recent years, block principal pivoting method [11, 12] has been developed that can speed up the search process by exchanging multiple variables instead of one variable between two working sets. Moreover, different versions of NTF methods, such as sparse NTF, discriminant NTF, and NTF on manifold, are also being introduced over the years [17, 24, 28]. In this article, a genetic algorithm-based method has been proposed that consistently provides better solutions across all values of K.

4 Proposed Genetic Algorithm-Based Technique

This section describes the use of genetic algorithm [19, 20] for evolving a near-optimal solution for matrix factorization. The proposed technique is described below in detail.

A genetic algorithm is an effective optimization algorithm based on the concept of Darwinian evolution. In this optimization technique, a population is generated randomly and it comprises a set of chromosomes. These chromosomes basically act as the candidate solutions, and it encodes the parameters of the search space. A fitness function is used to measure the quality of the solutions, and it is optimized iteratively. Genetic operators like crossover, selection, and mutation are employed for subsequent generations. Finally, it finishes its execution when the convergence criteria are met.

4.1 Encoding of Chromosomes

Suppose a matrix with dimension $M \times N$ is being factorized into two matrices W and H with dimensions $M \times K$ and $K \times N$, respectively. As these matrices are initialized randomly, therefore to encode the information of each solution into a chromosome, the length of it should be $M \times K + K \times N$. Every part of the chromosome is encoded with floating point values that represent the values of the matrices for a fixed division point K. Since this chromosome comprises the information about two matrices W and H with dimensions $M \times K$ and $K \times N$, respectively, so the first $M \times K$ positions of it represent the matrix W in row major, whereas the positions from $M \times K + 1$ to $M \times K + K \times N$ represent the matrix H in row major. Note that in this problem, we have considered that all the chromosomes have same length, i.e., the division point K is fixed for all the matrices. In this problem, primarily the matrix is normalized (to make the values between 0 and 1) that requires the cells of the chromosomes to be floating point values. The encoding scheme of the chromosome is shown in Fig. 1.

A sample representation of a chromosome for a matrix with dimension 10×15 is shown in Fig. 2. In this example, the decomposition point is chosen as 8. Therefore,

Fig. 1 Encoding scheme of chromosomes

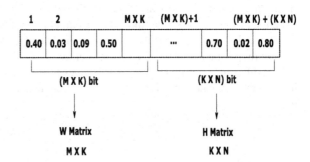

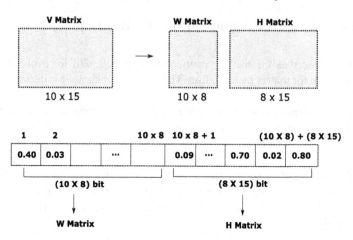

Fig. 2 An example of encoding scheme of chromosome containing the elements of a matrix with dimension 10×15. Here, the division point is chosen as 8

original matrix can be factorized into two matrices with order 10×8 and 8×15. To contain all these elements of the two matrices in a chromosome, the length of the chromosome should be of $10 \times 8 + 8 \times 15 = 200$ bits.

4.2 Initial Population

In the initial population, the whole set of solutions are generated by initializing two random matrices and those set of solutions representing different random matrices (with same dimensions) are encoded in the chromosomes. Thus, the initial population is created.

4.3 Local Refinement

In the subsequent step, a local refinement is performed in the chromosomes to accelerate the progress for finding the solutions quickly. In this phase, the chromosomes are revised by a little amount using gradient descent algorithm. The updation formula used to revise the elements is described below.

The difference (termed as the error) between the estimated rating ($\hat{v}$) and the original rating (v) is calculated by using the equation described below for each user-item pair.

$$e_{ij}^2 = (v_{ij} - \hat{v}_{ij})^2 = \left(v_{ij} - \sum_{k=1}^{K} w_{ik} h_{kj}\right)^2 \tag{3}$$

Now the elements of the chromosomes are revised by the following formulae.

$$w'_{ik} = w_{ik} + \alpha \frac{\delta}{\delta w_{ik}} e^2_{ij} = w_{ik} + \alpha(2e_{ij}h_{kj} - \beta w_{ik}) \tag{4}$$

$$h'_{kj} = h_{kj} + \alpha \frac{\delta}{\delta h_{kj}} e^2_{ij} = h_{kj} + \alpha(2e_{ij}w_{ik} - \beta h_{kj}) \tag{5}$$

Here, e_{ij} is the error generated in predicting (i,j)th element of the original matrix. w_{ik} denotes (i,k)th element of W matrix, and h_{kj} denotes the (k,j)th element of H matrix. w'_{ik} and h'_{kj} are the revised value of w_{ik} and h_{kj}, respectively. α is a constant used for controlling the convergence rate, and it is usually chosen as very small value; e.g., 0.0002. β is the regularization parameter, and it is set in the range of 0.02.

4.4 Objective Function

The prime objective here is to minimize the difference between the nonzero entries of the original matrix V and predicted matrix obtained after multiplying the two matrices W and H. Let us consider the dimensions of V, W, and H are $M \times N$, $M \times K$, and $K \times N$, respectively. Here, K is chosen an integer, and $K \leq min\{M, N\}$ as it is the rank of the matrix V. The distance between the nonzero cell values of the matrix is denoted by $||V - WH^T||^2$. The simple measure is termed as error due to the proposed solution.

On the other hand, to investigate the extent of linear relationship of respective cell values of both the matrices, correlation metric is used. Hence, a greater correlation value means there is a close relationship between the two matrices.

Thus, the final objective is to minimize the ratio

$$f = \frac{||V - WH^T||^2}{Corr(V, WH^T) + c}, \tag{6}$$

where $Corr(V, WH^T)$ denotes the correlation between two matrices V and WH. To avoid the indefinite condition for $Corr(V, WH^T) = 0$, a nominal value of $c = 1$ is added in the denominator. The goal is to minimize f.

4.5 Selection

The selection process chooses chromosomes for later breeding guided by the notion of "survival of the fittest" of natural genetic systems. In tournament selection strategy, it executes several "tournaments" among a few individuals that are chosen randomly from the initial population. In this context, the selection is based on binary tournament selection strategy, where the tournament size is 2.

4.6 Crossover

Crossover is a probabilistic process that interchanges information between two parent chromosomes to evolve two child chromosomes. In this article, crossover with a "fixed" crossover probability of k_c is used. In traditional genetic algorithm, crossover is normally done by single point or multipoint crossover. Here, multipoint crossover with a binary mask is performed.

4.7 Mutation

Each chromosome undergoes mutation with a very small mutation probability m_p. In the mutation operator, we have used to add some random float value ranging between 0–1 to each of the cell values of chromosome that is undergoing mutation.

4.8 Elitism

To preserve the best solution up to the current generation in the search space, the elitism method is applied. So the current best solution is retained for future generation with an aim to generate near-optimal solutions in minimal time.

5 Empirical Analysis

In this section, the datasets used for the experiments are discussed. Experiments have been performed on four artificial datasets and one real-life dataset (with different dimensions) to evaluate the performance of the proposed algorithm. The algorithm is compared with various well-known existing NMF methods, namely alternating least square non-negativity matrix with block principle pivoting (ANLS-BPP) [11, 12], alternating least square with active set method and column grouping (ANLS-ASGROUP) [11, 12], alternating least square (ANLS) [2], hierarchical least square (HALS) [7], multiplicative update method (MU) [15], and another EM-based method (LEE) [14]. The adopted performance metric is the squared error measure between the nonzero elements of original matrix and predicted matrix.

Experiments are performed in MATLAB 2008a, and the running environment is an Intel (R) CPU 1.6 GHz machine with 4 GB of RAM running Windows XP Professional.

5.1 Datasets

Four artificial datasets are used for experiments. We provide a short description of the datasets in terms of the dimension and the sparseness in Table 1. The artificial datasets are generated with uniformly distributed random values between 0 and 1. Then, some of the cells of this matrix are replaced by zeros that have a value less than a threshold value 0.5. Any threshold value (between 0 and 1) can be chosen to perform the discretization to create the missing entries. Hence, in this way, the artificial datasets with different dimensions are produced.

Two real-life gene expression datasets are used to perform the experiments. This dataset contains gene expression values for samples of prostate tumors. It contains 50 normal tissues and 52 prostate tumor samples. The expression matrix consists of

Table 1 Description of the artificial datasets

Dataset	Dimension	Sparsity (%)
Dataset 1	10×20	48.50
Dataset 2	20×30	47.33
Dataset 3	30×60	47.67
Dataset 4	50×100	55.16

Table 2 Description of the real-life dataset

Dataset	Dimension	Sparsity (%)
Dataset 5	102×44	50.98
Dataset 6	102×22	39.48

Table 3 Performance metric values for artificial data set of dimension 10×20 (Dataset 1)

Algorithm	$K = 2$	$K = 4$	$K = 6$	$K = 8$
ANLS-BPP	10.2864	4.7590	2.0972	1.0040
ANLS-ASGROUP	10.2889	4.7591	2.1044	0.8957
ALS	10.2587	4.5793	1.8561	1.8456
HALS	10.2876	4.7599	2.3495	0.8933
MU	10.2876	4.7599	2.3495	0.8939
LEE	10.4544	6.1094	3.7342	1.9383
Proposed	**5.0557**	**1.7189**	**1.0702**	**0.4543**

Best value in each column is given in bold

12,533 number of genes and 102 number of samples. It is publicly available in this website: http://www.biolab.si/supp/bi-cancer/projections/. For our current experimental purpose, we have arbitrarily taken two subsets of genes from the dataset. The first one contains 44 genes and 102 number of samples. Another one contains 22 genes with same number of samples. Some of the gene expression values have been made zeros randomly to make the missing entries. After that, the matrix is normalized (to make the values between 0 and 1) and experiment is performed on it.

Table 4 Performance metric values for artificial data set of dimension 20×30 (Dataset 2)

Algorithm	K = 8	K = 10	K = 12	K = 14
ANLS-BPP	13.3026	9.0797	6.9703	4.5536
ANLS-ASGROUP	13.4670	9.0735	7.0635	4.7477
ALS	11.5879	8.1087	7.3214	6.3251
HALS	13.1772	9.4310	7.0577	4.7590
MU	13.7973	9.7845	6.9314	4.5905
LEE	20.9117	18.5735	14.2420	12.9973
Proposed	**3.8492**	**3.1559**	**2.0468**	**1.4396**

Best value in each column is given in bold

Table 5 Performance metric values for artificial data set of dimension 30×60 (Dataset 3)

Algorithm	K = 10	K = 20	K = 24	K = 28
ANLS-BPP	58.0940	24.0208	12.46478	3.9299
ANLS-ASGROUP	58.1933	22.9045	12.2092	3.8016
ALS	48.0562	28.4527	25.6599	24.3488
HALS	58.1620	23.9880	27.2295	20.1077
MU	59.0062	24.9181	13.0354	4.2119
LEE	72.5032	51.0579	34.9671	24.1669
Proposed	**18.0383**	**5.9381**	**4.4476**	**3.7861**

Best value in each column is given in bold

Table 6 Performance metric values for artificial data set of dimension 50×100 (Dataset 4)

Algorithm	K = 10	K = 20	K = 30	K = 40
ANLS-BPP	206.5429	128.9246	75.6988	32.9996
ANLS-ASGROUP	204.1876	130.6368	76.2358	33.9127
ALS	172.9057	92.2291	72.6239	68.9306
HALS	206.4457	133.3246	99.0845	80.6341
MU	206.2046	128.5633	75.5515	37.0345
LEE	226.1495	176.9273	117.9271	64.2183
Proposed	**86.8549**	**47.2122**	**26.5956**	**18.4038**

Best value in each column is given in bold

A brief description of the dataset in terms of the dimension and the sparseness is provided in Table 2.

5.2 Parameter Settings

In this experimental setting, the length of the chromosomes is fixed for a particular dataset. For the proposed algorithm, the crossover rate is 0.9, mutation rate is 0.01, population size is 40, and number of generations is 400. All the parameters have been chosen empirically.

5.3 Comparative Results

We apply this algorithm to the four artificial and two real-life datasets, and the error values obtained by different matrix factorization methods along with the proposed method are shown in Tables 3, 4, 5, 6, 7, and 8, respectively. To check how this method performs for different values of K, we have varied K. It is easily observed that in almost all cases, the proposed matrix factorization method provides good perfor-

Table 7 Performance metric values for real-life data set of dimension 102×44 (Dataset 5)

Algorithm	$K = 4$	$K = 6$	$K = 10$	$K = 20$
ANLS-BPP	79.7494	56.4053	56.5671	32.8333
ANLS-ASGROUP	79.5560	70.4775	56.5108	33.7389
ALS	76.8062	63.8603	44.0499	20.7839
HALS	79.9604	70.1249	56.9198	40.8811
MU	70.9779	70.7872	56.8998	35.3102
LEE	81.2448	73.9244	69.3997	52.5873
Proposed	**6.5560**	**5.8166**	**5.2323**	**3.7925**

Best value in each column is given in bold

Table 8 Performance metric values for real-life data set of dimension 102×22 (Dataset 6)

Algorithm	$K = 2$	$K = 5$	$K = 7$	$K = 10$
ANLS-BPP	21.3561	14.1517	11.5843	8.5341
ANLS-ASGROUP	21.3509	11.6273	56.5108	8.7800
ALS	21.2720	13.8371	10.5881	7.0037
HALS	21.3536	14.3428	12.5611	8.6388
MU	21.3574	14.5042	12.0127	8.9663
LEE	23.0770	18.0111	16.3493	14.6988
Proposed	**5.3391**	**4.4688**	**4.3779**	**3.6325**

Best value in each column is given in bold

Fig. 3 Generation-wise
fitness value for dataset 1

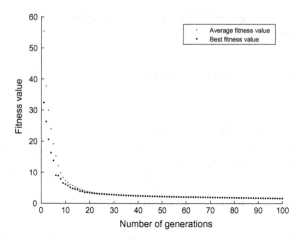

Fig. 4 Generation-wise
fitness value for dataset 2

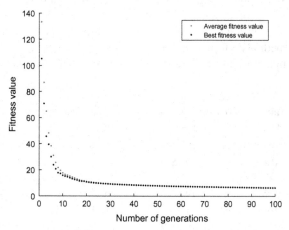

Fig. 5 Generation-wise
fitness value for dataset 3

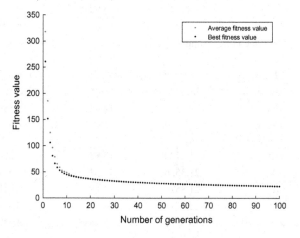

Fig. 6 Generation-wise fitness value for dataset 4

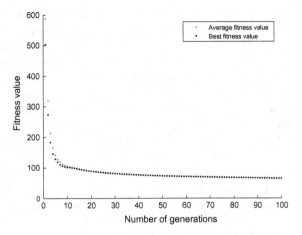

Fig. 7 Generation-wise fitness value for dataset 5

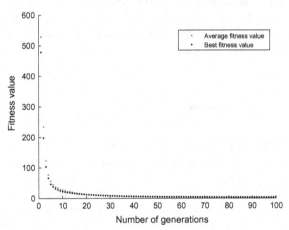

Fig. 8 Generation-wise fitness value for dataset 6

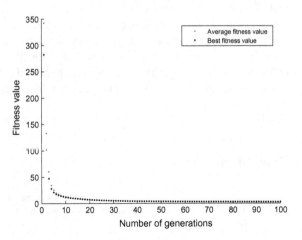

mance consistently. Moreover, it performs better in all K values, which demonstrates the utility of the evolutionary framework for developing such type of methods.

It is observed that for initial value of K, the proposed method is far better than the other methods. The margins of differences decrease if the value of K is increased further. Again, it is easily seen that the proposed approach is consistently producing the best result for most of the K values, and it generates more accurate result if the number of generations is allowed to be increased further. In Figs. 3, 4, 5, 6, 7, and 8, the generation-wise best fitness values (shown in black) and average fitness values (shown in red) produced by the proposed method are plotted. It is noticed that the best fitness values over the generations improve gradually in a steady-state manner. Again as elitism is incorporated, there is always an improvement of the best fitness value over the generations. From these tables, it can be observed that in some of the situations, ANLS-BPP method is providing competitively close result. As example, for dataset 4 (at $K = 40$) and dataset 5 (at $K = 6$), it produces competitively closer result. With respect to all the methods, it is realized that the performance of ANLS-BPP, ANLS-ASGROUP, and HALS is competitively close to that of the proposed method in few cases. In case of real-life datasets, ALS method performs better in most of the K values.

6   Conclusion

In this article, to mitigate the common sparsity problem faced in the bioinformatics, crowdsourcing, and recommender system, a matrix factorization technique using evolutionary algorithm is proposed. In this context, the objectives are to minimize the distance between the nonnegative elements of the original matrix and the predicted matrix, whereas maximizing the correlations of the nonzero elements between the two matrices. The performance of the proposed method has been compared with that of other existing matrix factorization methods on some artificial datasets and a real-life gene expression dataset. The effectiveness of the proposed technique over other existing approaches is easily realized from the experimental results. In future study, this method can be extended in multi-objective evolutionary framework to optimize both the objective functions simultaneously. Moreover, other conflicting objective functions can be used, and this technique can be very beneficial to make better prediction.

References

1. Adomavicius, G., Tuzhilin, A.: Towards the next generation of recommender systems: a survey of the state-of-the-art and possible extensions. IEEE Trans. Knowl. Data Eng. **6**, 734–749 (2005)

2. Berry, M., Browne, M., Langville, A., Pauca, V., Plemmons, R.: Algorithms and applications for approximate nonnegative matrix factorization. Comput. Stat. Data Anal. **52**(2), 155–173 (2007)
3. Brabham, D.C.: Detecting stable clusters using principal component analysis. Methods Mol. Biol. **224**(10) (2013)
4. Brunet, J., Tamayo, P., Golub, T., Mesirov, J.: Metagenes and molecular pattern discovery using matrix factorization. Proc. Nat. Acad. Sci. **101**, 1464–1469 (2004)
5. Chatterjee, S., Mukhopadhyay, A.: A genetic algorithm-based matrix factorization in missing value prediction. In: Proceedings of the 1st International Conference on Computational Intelligence, Communications, and Business Analytics. Springer, CCIS, Kolkata, India (2017)
6. Christidis, K., Mentzas, G.: A topic-based recommender system for electronic marketplace platforms. Expert Syst. Appl. **40**(11), 4370–4379 (2013)
7. Cichocki, A., Zdunek, R., Amari, S.I.: Hierarchical als algorithms for nonnegative matrix and 3d tensor factorization. In: Lecture Notes in Computer Science, vol. 4666, pp. 169–176. Springer (2007)
8. Demartini, G., Difallah, D.E., Mauroax, C.: Zencrowd: leveraging probabilistic reasoning and crowdsourcing techniques for large scale entity linking. In: Proceedings of the 21st International Conference on World Wide Web, pp. 469–478. Lyon, France (2012)
9. Ding, C.H.Q., Li, T., Jordan, M.I.: Convex and semi-nonnegative matrix factorizations. IEEE Trans. Pattern Anal. Mach. Intell. **32**(1), 45–55 (2010)
10. Friedman, A., Berkovsky, S., Kaafar, M.A.: A differential privacy framework for matrix factorization recommender systems. User Model. User-Adap. Interact. **26**(5), 425–458 (2016)
11. Kim, J., He, Y., Park, H.: Algorithms for nonnegative matrix and tensor factorizations: a unified view based on block coordinate descent framework. J. Global Opt. **58**(2), 285–319 (2014)
12. Kim, J., Park, H.: Fast nonnegative matrix factorization: an active-set-like method and comparisons. SIAM J. Sci. Comput. (SISC) **33**(6), 3261–3281 (2011)
13. Kittur, A., Nickerson, J.V., Bernstein, M., Gerber, E., Shaw, A., Zimmerman, J., Lease, M., Horton, J.: The future of crowd work. In: Proceedings of the CSCW, pp. 1301–1318 (2013)
14. Lee, D., Seung, H.: Learning the parts of objects by non-negative matrix factorization. Nature **401**, 788–791 (1999)
15. Lee, D., Seung, H.: Algorithms for non-negative matrix factorization. Proc. Adv. Neural Inf. Process. Syst. **13**, 556–562 (2001)
16. Lin, C.J.: Projected gradient methods for nonnegative matrix factorization. Neural Comput. **19**(10), 2756–2779 (2007)
17. Liu, J., Liu, J., Wonka, P., Ye, J.: Sparse non-negative tensor factorization using columnwise coordinate descent. Pattern Recognit. **45**(1), 649–656 (2012)
18. Luo, X., Liu, H., Gou, G., Xia, Y., Zhu, Q.: A parallel matrix factorization based recommender by alternating stochastic gradient decent. Eng. Appl. Artif. Intell. **25**(7), 1403–1412 (2012)
19. Maulik, U., Bandyopadhyay, S.: Genetic algorithm based clustering technique. Pattern Recognit. **32**, 1455–1465 (2000)
20. Mukhopadhyay, A., Maulik, U., Bandyopadhyay, S.: Multiobjective genetic fuzzy clustering of categorical attributes. In: Proceedings 10th International Conference on Information Technology, pp. 74–79 (2007)
21. Mukhopadhyay, A., Maulik, U., Bandyopadhyay, S.: A survey of multiobjective evolutionary clustering. ACM Comput. Surv. **47**(4), 61:1–61:46 (2015)
22. Paatero, P.: Least squares formulation of robust non-negative factor analysis. Chem. Intell. Lab. Syst. **37**(1), 23–35 (1997)
23. Paatero, P., Tapper, U.: Positive matrix factorization: a non-negative factor model with optimal utilization of error estimates of data value. Environmetrics **5**(2), 111–126 (1994)
24. Phan, A.H., Tichavský, P., Cichocki, A.: Fast damped gauss-newton algorithm for sparse and nonnegative tensor factorization. In: Proceedings of the IEEE International Conference on Acoustics, Speech, and Signal Processing, ICASSP, pp. 1988–1991. Prague, Czech Republic (2011)

25. Ross, J., Irani, L., Silberman, M., Zaldivar, A., Tomilson, B.: Who are the crowdworkers? shifting demographics in mechanical turk. In: Proceedings of the SIGCHI Conference on Human Factors in Computing Systems, pp. 2863–2872 (2010)
26. Xu, W., Liu, X., Gong, Y.: Document clustering based on non-negative matrix factorization. In: Proceedings of the 26th Annual International ACM SIGIR Conference on Research and Development in Information Retrieval, vol. 101, pp. 267–273. ACM Press (2003)
27. Yoo, J., Choi, S.: Orthogonal nonnegative matrix tri-factorization for co-clustering: multiplicative updates on stiefel manifolds. Inf. Process. Manage. **46**(5), 559–570 (2010)
28. Zafeiriou, S.: Algorithms for Nonnegative Tensor Factorization, pp. 105–124. Springer, London (2009)

Differential Evolution in PFCM Clustering for Energy Efficient Cooperative Spectrum Sensing

Anal Paul and Santi P. Maity

Abstract Cooperative spectrum sensing (CSS) in cognitive radio network (CRN) is highly recommended to avoid the interference from secondary users (SUs) to primary user (PU). Several studies report that clustering-based CSS technique improves the system performance, among them fuzzy c-means (FCM) clustering algorithm is widely explored. However, it is observed that FCM generates an improper clustering of sensing information at low signal-to-noise ratio (SNR) due to inseparable nature of energy data set. To address this problem, the present chapter describes a work that investigates the scope of possibilistic fuzzy c-means (PFCM) algorithm on energy detection-based CSS. PFCM integrates the possibilistic information and fuzzy membership values of input data in the clustering process to segregate the indistinguishable energy data into the respective clusters. Differential evolution (DE) algorithm is applied with PFCM to maximize the probability of PU detection (P_D) under the constraint of a target false alarm probability (P_{fa}). The present work also evaluates the required power consumption during CSS by SUs. The proposed technique improves P_D by $\sim 12.53\%$ and decreases average energy consumption by $\sim 5.34\%$ over the existing work.

Keywords Spectrum sensing · Fuzzy c-means clustering
Possibilistic fuzzy c-means clustering · Differential evolution algorithm

1 Introduction

Cognitive radio (CR) is a promising technology to overcome the spectrum scarcity problem as CR exploits the ideal radio band [1–6] to offer an opportunistic data transmission mode. In CR networks (CRNs), the unlicensed users are known as

A. Paul · S. P. Maity (✉)
Department of Information Technology, Indian Institute of Engineering
Science and Technology, Shibpur, Howrah 711103, India
e-mail: santipmaity@it.iiests.ac.in

A. Paul
e-mail: ap.rs2015@it.iiests.ac.in

© Springer Nature Singapore Pte Ltd. 2019
J. K. Mandal et al. (eds.), *Advances in Intelligent Computing*,
Studies in Computational Intelligence 687,
https://doi.org/10.1007/978-981-10-8974-9_5

secondary users (SUs) or cognitive users (CUs) and users licensed with spectrum are called as primary users (PUs). Spectrum sensing (SS) is an important part in CRN to avoid harmful interference with PUs, and it identifies the available spectrum for opportunistic data transmission [4]. The reliability of SS is estimated by high value of probability of PU detection (P_D) and low value of probability of false alarm (P_{fa}). In cooperative SS (CSS), geographically distributed several SUs sense a particular frequency band and forward the sensing information to the fusion center (FC) for obtaining a global decision. A suitable CSS scheme mitigates the shadowing and fading effects, multipath problem, hidden node issue, etc., over non-CSS technique [6] and provides a high P_D accuracy with reliability [7].

Performance evaluation of CSS with different techniques is widely investigated and reported in the literature [8–17], such as energy detection (ED) [8–10], cyclostationary feature detection [11], matched filtering [12], eigenvalue-based detection [13], entropy-based sensing [14], wavelet-based detection [15], generalized likelihood ratio test [16], waveform-based detection [17], etc. Several pros and cons of the above-mentioned schemes are also reported [18, 19]. However, ED-based CSS technique is widely used in practice due to its low computational cost and implementation simplicity [18, 20].

Another concern regarding CSS is energy consumption by SUs during the transmissions of sensing information (i.e., amplify-and-forward) to FC. Several energy efficient techniques are reported in the literature [21–27]. Participation of several SUs in SS improves the CSS performance but it also increases the SU energy consumption [26, 27]. The overall energy consumption during CSS is depended on multiple parameters, such as required number of samples (i.e., sensing period), number of SUs involvements in SS, associated amplifying power gain for transmitting the sensing information to FC [23]. Though some amount of power is consumed in the processing circuit of an amplify-and-forward (AF) relay acting as sensing node for sensing the PU samples, however, this amount is much lower compared to the transmission power consumption. In other words, a significant amount power is consumed in the amplifying process. Therefore, a trade-off between the sensing reliability and the energy consumption during CSS exists [26, 27]. The accurate calculation of amplifying gain for individual SU from a large set of SUs is always intractable as several parameters are involved in complex optimization techniques or sometimes it also suffers from lack of availability of closed form expression [25, 26]. Hence, a simple algorithm with low computational complexity that even may provide a sub-optimal solution becomes a preferable choice. It is found that cluster-based techniques are found to be efficient to offer solutions [22, 25, 26, 28].

The rest of this chapter is arranged in the following ways: Literature study, limitations, significant contributions of the present work are discussed in Sect. 2. The typical system model for the proposed work is explained in Sect. 3. The PFCM clustering and differential evolution (DE) optimization technique are combined in Sect. 4. A set of simulation results and numerical analysis are demonstrated in Sect. 5. At the end, the chapter is concluded in Sect. 6.

2 Literature Review

The scope of cluster-based approaches followed by differential evaluation (DE) optimization strategy in CSS is discussed in this section.

2.1 Cluster-Based Approaches

Several clustering algorithms are exploited in CSS to enhance the performance of CRN [22, 24–26, 28]. The authors [28] applied FCM clustering on the received energy vales at FC and partitioned the input data space into four different clusters [28]. It is observed that FCM-based CSS technique improved the SS performance. However, the work did not investigate the actual energy consummation during CSS [28]. The further study [26] reported the energy efficient CSS approach. FCM clustering technique is combined with DE algorithm to obtain the minimum amplifying gain along with an optimal number of PU samples while sensing constraints were imposed. This strategy outperformed the traditional energy efficient algorithm [26, 27]. However, FCM performance degrades at low SNR due to the presence of excessive noise in data set leading to a non-spherical nature of input data space formed from the energy set at FC [25, 29]. In FCM, the membership value of a data point is calculated using the Euclidean distance from that particular point to the clusters centers. A data point having equal distances from the multiple cluster center has the same membership values for all the clusters which leads to an improper clustering. The objective of the typical clustering technique is to produce the clusters with low intra-cluster distances/(compactness) and high inter-cluster distances.

The authors [25] applied the kernel FCM (KFCM) clustering in CSS technique to overcome the drawbacks of FCM at low SNR. KFCM applies the specific kernel function to project the inseparable data from input space onto a high-dimensional space. This projection helps a linear classifier to separate the data in the respective clusters [29]. KFCM-based CSS [25] extended CRN performance and reduced power consumption compared to the other techniques [25, 27, 28]. KFCM selects a best kernel from the set of kernels, and this selection is highly influenced by the input patterns of data [30]. Thus, KFCM suffers from the high sensitivity to select the appropriate kernel, and any improper selection may lead to the local optimal issues [30, 31]. The use of fuzzy-possibilistic c-means (FPCM) clustering solves the sensitivity issue of KFCM but FPCM encounters some challenges during the updation of the typicality and membership values if the input data are unlabelled [32–34]. Possibilistic c-means (PCM) algorithm solves the cluster overlapping problem of FCM and value updating issues of FPCM but it suffers from stability problems [32].

2.2 Differential Evolution-Based Optimization in CRN

DE is a stochastic evolutionary algorithm and is used in several problems for obtaining the global optimal solution [35–38]. DE adopts the quick searching technique that involves multiple iteration to enhance the fitness of the candidate solution. It attempts to optimize the global solution by applying mutation and crossover to generate the new candidate solution. The fitness of the new solutions determines the replacement possibility of the existing population members [35]. One may also suggest to use other evolutionary algorithms (EAs) like genetic algorithms (GAs), evolution strategies (ESs), biogeography-based optimization (BBO), evolutionary programming (EP), ant colony optimization (ACO), genetic programming (GP), cultural algorithm (CA), particle swarm optimization (PSO) for such random search. However, in the proposed work the calculated energy values of the received samples are in the form of real numbers. It is worth mentioning that DE provides better system stability and control over the typical GA where the set of initial population and chromosomes are in real numbers. Hence, DE is used instead of GA in the present work. On the contrary, unlike the other EAs and ESs, DE computes the differences from the evolving parameter vectors in a particular sequence to explore its objective function. In many real-time implementations, EAs and ESs may encounter some optimization challenges such as unexpected changes in the environmental parameters' values after optimization, uncertainty in the noisy fitness function during the genetical evolutions, aberration of global optimal point after the termination of the searching process, approximation errors in fitness function. All these problems of EAs and ESs are efficiently tackled by DE algorithm, as DE inherits the complex searching policy in the large continuous spaces and offers an extensive performance evaluation in terms of convergence speed, global optimal accuracy, and robustness in linear or in nonlinear optimization problems. The effectiveness of DE algorithm is also investigated in the field of CRN to achieve SS reliability, proper spectral allocation, and in optimal power control strategies [25, 26, 39–43]. DE scheme, associated with the several clustering schemes (i.e., FCM and KFCM), is used for finding the optimal cluster points which in turn increase the CSS performance and minimize the overall power consumption by the SUs [25, 26].

2.3 Scope and Contributions

To subdue the several drawbacks of different clustering techniques, the proposed work investigates the possibilistic fuzzy c-means (PFCM) clustering algorithm at low SNR. It is worth mentioning that PFCM solves the noise sensitivity issues of FCM, provides better stability over PCM, removes the row aggregation constraint of FPCM [32], and eliminates the kernel selection issues of KFCM [34]. PFCM determines the typicality and membership values of unlabelled data simultaneously throughout the clustering process. PFCM is a well-defined combination of PCM and FCM techniques

[32]. PFCM clustering along with DE is applied in the proposed work to obtain the minimal SU amplifying gain to reduce the average power consumption. The proposed work considers that SUs are spatially distributed and a set of associative SUs forms a cluster using PFCM. An optimal amplifying power gain of individual SU cluster is computed through the DE algorithm. The notable findings of the present works are as follows:

- Inseparable energy values are segregated into respective clusters using PFCM. The proposed approach offers better CSS performance over the other existing works in [25, 26, 28].
- DE optimization strategy is combined with PFCM clustering policy to obtain an optimal amplifying gain and a minimum number of sensing samples for reducing the average energy consumption during CSS.
- An optimal amplifying power gain for each SU cluster is determined, instead of same power allocation strategy among all the SUs [27].
- The substantial set of simulation results along with numerical analysis demonstrates the superiority of the proposed work in terms of PU detection and optimal energy consumption compared to the existing works [25–27].

3 System Model

Figure 1a represents the typical system model of the proposed work. It contains a PU transmitter, K number of secondary transceivers (i.e., SUs), and an FC. FC consists of K number of antennae $(AT_1, AT_2, AT_3, \ldots, AT_K)$. SUs are spatially distributed, and each SU requires an amplifying gain for transmitting the sensed samples to FC. The computation of an amplifying gain for individual SU is impractical as optimization complication is exponentially increased with the increment in the number of SUs [27]. This problem is solved by the cluster based approaches where individual SU cluster uses a single amplifying gain [25, 26]. A set of associative SUs form a cluster by depending on the distances between PU-SU (d_{aK}) and SU-FC (d_{bK}). All sensing channels are assumed to follow the Rayleigh distribution, and the reporting channels are considered to be ideal. The path loss exponent (α) of the channel is considered to be random for different channel states.

The typical time frame structure for SS is depicted in Fig. 1b. The sensing interval (T_{sense}) is partitioned into several identical sensing slots (t_s), and individual t_s is further subdivided into two equal sub-slots (i.e., t_{s1} and t_{s2}). During t_{s1}, SUs sense the primary signal, and during t_{s2}, SUs amplify and forward the received energy samples to FC using an amplifying gain factor $\sqrt{\omega_i}$, $\forall i \in \{1, 2, \ldots, K\}$. PU is either busy or ideal on a particular instant over a specific frequency band, and it can be estimated from the received energy samples at FC. The binary hypotheses $\mathcal{H}_1$ and $\mathcal{H}_0$ represent the active and ideal states of PU, respectively.

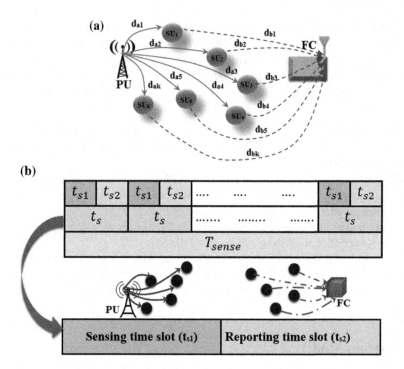

Fig. 1 a Proposed system model and b typical time frame structure

3.1 Signal Model

During t_{s1}, the obtained primary signal (R_i) at ith SU node is expressed as

$$R_i(t) = \eth\sqrt{P_{pow}}\,\hbar_i\,Y_i(t) + \varnothing_i(t), \ \forall t \ \in [1, N] \tag{1}$$

The binary variable '$\eth$' indicates the presence of PU ($\eth = 1$) or the absence ($\eth = 0$) on a particular instant.

$$\mathcal{H}_1 : R_i(t) = \sqrt{P_{pow}}\,\hbar_i\,Y_i(t) + \varnothing_i(t),$$
$$\mathcal{H}_0 : R_i(t) = \varnothing_i(t) \tag{2}$$

The symbol P_{pow} denotes the transmission power of PU, and $\hbar_i$ represents the channel fading coefficient between PU-SU$_i$ where $h_i(t) \sim CN(0, d_{a_i}^{-\alpha})$. It is assumed that the distribution of PU signal $Y_i(t)$ is circularly symmetric complex Gaussian (CSCG) with zero mean and variance is σ_p.

$$\sigma_p = E[|Y_i(t)|^2] = P_{pow} \tag{3}$$

The symbol $\varnothing_i(t)$ represents the noise at ith SU, and it is independent and identically distributed (IID) CSCG with zero mean and variance is σ_n.

$$\sigma_n = E[|\varnothing_i(t)|^2] = P_{noise} \tag{4}$$

All SUs forward the sensed samples to FC. Therefore, the accumulated signal at FC is expressed as

$$Z_i(t) = \sum_{i=1}^{K} \sqrt{\omega_i} \hbar_{ix} R_i(t) + \varnothing_x(t) \, \forall \, t \, \in [1, N] \tag{5}$$

The variables '$\hbar_{ix}$' and '$\varnothing_x$' indicate the channel fading coefficient between $(SU_i$-$FC)$ and noise at FC, respectively. The acquired energy value (E_i) of all received samples at ith SU is determined as

$$E_i = \sum_{n=1}^{N} |(Z_i(t))|^2 \tag{6}$$

FC computes E_i values for the K number of SUs and forms an energy set $E_{SU_{All}} = \{E_1, E_2, E_3 \dots E_i\}_{i=1}^{K}$. The maximum E_i value is selected by FC $(max\{E_{SU_{All}}\})$ and stores in $\{E_{Max_{final}}\}$ to obtain the final energy set. FC continuously includes $max\{E_{SU_{All}}\} \longmapsto \{E_{Max_{final}}\}$ for different channel states during SS.

3.2 Consumption of Energy During SS

The power consumption $(Power_i)$ at ith SU is calculated as [26]

$$Power_i = (\mathcal{P}(\mathcal{H}_1)\omega_i((d_a)_i^{-\alpha}P_{pow} + P_{noise}) + \mathcal{P}(\mathcal{H}_0)\omega_i P_{noise}) \tag{7}$$

The symbols '$\mathcal{P}(\mathcal{H}_1)$' and '$\mathcal{P}(\mathcal{H}_0)$' indicate the probability of $\eth = 1$ and $\eth = 0$, respectively. The distinct amplifying gain for individual SU cluster is determined as [25, 26]

$$\omega_c = \sum_{i=1}^{c} \omega_{ci} \tag{8}$$

The symbol 'ω_{ci}' indicates the required amplifying gain of ith cluster. Power gain of kth SU at ith cluster is derived as

$$\omega_{c_{iK}} = \frac{d_{a_{iK}}}{\sum_{i=1}^{K} d_{a_{iK}}} \tag{9}$$

The symbol '$d_{a_{i_K}}$' denotes the Euclidean distance from PU to SU_k.

The symbol 'E_{cons}' represents the average energy consumption of all SUs and is calculated as

$$E_{cons} = EN_{bp} + EN_{pa} + EN_c \tag{10}$$

where 'EN_{bp}' and 'EN_{pa}' indicate the energy consumption in baseband circuits and the energy required to supply the amplifying gain, respectively. The energy consumed during the signal reception and transmission at SUs is denoted as EN_c. It is observed that $EN_{pa} \gg (EN_{bp} \text{ and } EN_c)$, thus EN_{bp} and EN_c are ignored, and therefore,

$$E_{cons} \equiv EN_{pa} \tag{11}$$

$$EN_{pa} = \sum_{i=1}^{K} Power_i N(t_{sd}) \tag{12}$$

The symbol t_{sd} indicates the individual sample duration and $T_{sense} = N \times t_{sd}$. Using the specific ω_c and N values, the set of cluster centers is obtained through PFCM. The corresponding values of ω_c and N are selected into DE matting pool if the constraints of P_D and P_{fa} are satisfied. This process is iteratively repeated for all the possible combinations of ω_c and N.

To improve the readability of the proposed work, a list of symbols are included in Table 1.

4 PFCM-Based CSS Technique

FCM algorithm partitions the unlabelled data space $\mathcal{X}_D = \{x_1, x_2, \ldots, x_k, x_{k+1}, \ldots, x_{\mathcal{Z}}\}$ into C number of clusters using a fuzzy membership value. The fuzzy partition matrix $\cup$ of $\mathcal{X}_D$ is denoted as $\{A_i; 1 \leq i \leq C\} \subset P(\mathcal{X}_D)$. The membership value of x_k into ith cluster is represented by $\mathbb{U}_{ik} \in (0, 1)$. The matrix elements of $[c \times \mathcal{Z}]$ need to satisfy the constraints of [44]

$$C1: \mathbb{U}_{ik} \in [0, 1]; \ 1 \leq i \leq C, \ 1 \leq k \leq \mathcal{Z}$$

$$C2: \sum_{i=1}^{C} \mathbb{U}_{ik} = 1; \ k \subseteq \mathcal{X}_D, \ 1 \leq i \leq C$$

$$C3: 0 \leq \sum_{k=1}^{\mathcal{Z}} \mathbb{U}_{ik} \leq \mathcal{Z}; \ 1 \leq k \leq \mathcal{Z} \tag{13}$$

The possibilistic partition is obtained by altering the constraint C2 in Eq. (13) [30, 44] and redefined as

Table 1 A list of important symbols

Symbol	Interpretation
K	Number of participated SUs in sensing/diversity number
T_{sense}	Interval for the spectrum sensing
t_{s1}	Allocated time slot for spectrum sensing
$\eth = 1, \eth = 0$	PU presence, PU absence
t_{s2}	Reporting time slot for the SUs
R_i	Received PU signal at SU_i
Y_i	PU transmitted signal
$Z(t)$	Received signals at FC from SU
P_{pow}	Transmission power of the primary user
N	Count of PU samples
h_i	Channel fading coefficient between PU and SU_i
d_{aK}	Distance between PU transmitter and SU
d_{bK}	Distance between SU and FC
$\mathcal{H}_1$	Binary hypothesis when PU is present
$\mathcal{H}_0$	Binary hypothesis when PU is absent
$\mathcal{P}(\mathcal{H}_1), \mathcal{P}(\mathcal{H}_0)$	Probability of PU being active and idle
α	Path loss exponent
$\varnothing_i$	SU_i receiver noise
P_{noise}	Noise variance/power
E_i	Energy value of the acquired signal at SU_i
E_{max}	Strongest signal energy value at FC (using SC)
ω_{ci}	Power gain of ith cluster
$\omega_{c_{iK}}$	Amplifying power gain of kth SU $\in i$th cluster
ω_c	Sum amplifying gain of the SUs
$Power_i$	Power consumption at SU_i
E_{cons}	Average consumed energy by SUs
EN_{bp}	Energy consumption at SU circuits
EN_{pa}	Required amplification energy at SU_i to transmit the energy values at FC
EN_c	Consumed energy at SU during reception (PU-SU) and transmission (SU-FC) of signal
t_{sd}	Sample duration
$\mathcal{X}_\mathcal{D}$	Unlabelled input data space
c	Number of clusters
m	Fuzzifier, where $m > 1$
$\cup$	Fuzzy partition
$\mathbb{U}_{ik}$	Membership value of a data (x_k) in ith cluster
$\mathcal{V}$	$v_1, v_2, \ldots, v_c$ represents set of cluster centers
$T =$	Typicality matrix

(continued)

Table 1 (continued)

Symbol	Interpretation
τ_{ik}	Typicality value of the data element of input space
λ_{ikA}^2	Squared inner-product norm
W	Weighting exponent
γ_i	Distribution possibility
δ	Fuzzy membership
Φ	Typicality values
v_i	ith cluster center
A	Norm convincing matrix
Np^*	Initial population for DE
$\{v_1, v_2, v_3, v_4\}$	Four different cluster centers
$\mathcal{V}$	A set of all cluster centers
$\{\mathcal{V}_1^*, \mathcal{V}_2^*, \mathcal{V}_3^*, \mathcal{V}_4^*\}$	Four different optimal cluster centers
C_r	Crossover rate
$\hat{\mathcal{V}}$	The set satisfies $P_{fa} \leq P_{fa}^{th}$ in initial population
F	Scaling factor in DE
P_D	Probability of PU detection
P_{fa}	Probability of false alarm
P_{fa}^{th}	Probability of false alarm threshold

$$C1 : \mathbb{U}_{ik} \in [0, 1]; \ 1 \leq k \leq \mathcal{Z}, \ 1 \leq i \leq c$$
$$C2 : \exists i, \mathbb{U}_{ik} > 0, \ \forall k; \ 1 \leq i \leq c$$
$$C3 : 0 \leq \sum_{k=1}^{\mathcal{Z}} \mathbb{U}_{ik} \leq \mathcal{Z}; \ 1 \leq k \leq \mathcal{Z} \tag{14}$$

FCM optimization function is

$$\mathbb{F}(\mathbb{U}, \mathcal{V}) = \sum_{k=1}^{\mathcal{Z}} \sum_{j=1}^{c} (\mathbb{U}_{ik})^m \|x_k - v_i\|^2, \ where \ m > 1 \tag{15}$$

where 'v_i' and 'm' denote the ith cluster center and the fuzzifier, respectively. The symbol '$\mathcal{V}$' represents a set of cluster centers $v_1, v_2, \ldots, v_c$. The new cluster centers and the partition matrix are optimized and updated as [32]

$$v_i = \frac{\sum_{k=1}^{\mathcal{Z}} (\mathbb{U}_{ik})^m x_k}{\sum_{k=1}^{\mathcal{Z}} (\mathbb{U}_{ik})^m} \tag{16}$$

$$\mathbb{U}_{ik} = \left[\sum_{j=1}^{c} \left(\frac{||x_k - v_i||^2}{||x_k - v_j||^2} \right)^{\frac{2}{(m-1)}} \right]^{-1} \tag{17}$$

FPCM solves the noise sensitivity issues of FCM by simultaneously computing the typicality and the membership values. The objective function of FPCM is expressed as [32]

$$\mathbb{F}(T, \mathbb{U}, \mathcal{V}; \mathcal{X}_{\mathcal{D}}) = \sum_{j=1}^{c} \sum_{k=1}^{\mathcal{Z}} (\tau_{ik})^W \lambda_{ikA}^2 + \sum_{j=1}^{c} \sum_{k=1}^{\mathcal{Z}} (\mathbb{U}_{ik})^m \tag{18}$$

where the norm conclusive matrix [32] is represented by symbol 'A'. The typicality matrix (T) of input data element is denoted by $[\tau_{ik}]$. The λ_{ikA}^2 is defined as

$$\lambda_{ikA}^2 = ||x_k v_i||_A^2 = (x_k v_i)^T \tag{19}$$

The symbol 'W' represents the weighting exponent and $W > 1$ in (18). It determines the typicality values of the clusters ($0 \leq \mathbb{U}_{ik}, \tau_{ik} \leq 1$). FPCM has the following constraints: $\forall k$; $\sum_{k=1}^{\mathcal{Z}} \tau_{ik} = 1, \sum_{i=1}^{c} \mathbb{U}_{ik} = 1, \forall i$. If $\lambda_{ikA}^2 > 0, \forall k, i; W > 1, m > 1$, then $(T, \mathbb{U}, \mathcal{V}) \in \mathcal{R}^{n \times c}$ optimize as the following

$$\mathbb{U}_{ik} = \frac{1}{\sum_{j=1}^{c} (\lambda_{ikA}/\lambda_{jkA})^{2/(m-1)}}, \quad \forall i, k \tag{20}$$

$$\tau_{ik} = \frac{1}{\sum_{j=1}^{c} (\lambda_{ikA}/\lambda_{jkA})^{2/(W-1)}}, \quad \forall i, k \tag{21}$$

$$v_i = \frac{\sum_{k=1}^{\mathcal{Z}} ((\mathbb{U}_{ik})^m + (\tau_{ik})^W) x_k}{\sum_{k=1}^{\mathcal{Z}} ((\mathbb{U}_{ik})^m + (\tau_{ik})^W)}, \quad \forall i \tag{22}$$

FPCM provides a very small vale in the typicality matrix for a large data set. This problem is solved through a scale up [32]. However, this scaled parameter does not contribute much information regarding the data. PFCM solves this problem by relaxing the constraint on typicality matrix but holds the column constraint on $\mathbb{U}$. The objective function of PFCM is expressed as

88 A. Paul and S. P. Maity

$$\mathbb{F}(T, \cup, \mathcal{V}; \mathcal{X}_D) = \sum_{i=1}^{c} \gamma_i \sum_{k=1}^{z} (1 - \tau_{ik})^W + \sum_{i=1}^{c} \sum_{k=1}^{z} \left[(\delta \mathbb{U}_{ik})^m + (\Phi \tau_{ik})^W \right] \lambda_{ikA}^2 \quad (23)$$

while $m > 1, W > 1, \tau_{ik} \le 1, \gamma_i > 0, \sum_{i=1}^{c} \mathbb{U}_{ik} = 1, 0 \le \mathbb{U}_{ik}.$ Where γ_i determines the distribution possibility . The symbols Φ and δ indicate the typicality and fuzzy membership values, respectively. The values are usually chosen by depending upon the initial partition derived through FCM.

$$\gamma_i = \frac{\sum_{k=1}^{z} (\mathbb{U}_{ik})^W \lambda_{ikA}^2}{\sum_{k=1}^{z} (\mathbb{U}_{ik}^W)} \quad (24)$$

If $\lambda_{ikA}^2 > 0, \forall k, i; \ W > 1, \ m > 1$, and $(T, \cup, \mathcal{V}) \in \mathcal{R}^{n \times c}$, then

$$\mathbb{U}_{ik} = \frac{1}{\sum_{j=1}^{c} (\lambda_{ikA}/\lambda_{jkA})^{2/(m-1)}}, \ \forall k, i \quad (25)$$

$$v_i = \frac{\sum_{k=1}^{z} ((\delta \mathbb{U}_{ik})^m + (\Phi \tau_{ik})^W) x_k}{\sum_{k=1}^{z} ((\delta \mathbb{U}_{ik})^m + (\Phi \tau_{ik})^W)}, \ \forall i \quad (26)$$

$$\tau_{ik} = \frac{1}{1 + \left(\frac{\Phi}{\gamma_i} \lambda_{ikA}^2\right)^{1/(W-1)}}, \ \forall i, k \quad (27)$$

PFCM partitions the energy values into four different clusters at FC. The clusters $c_1, c_2 c_3$, and c_4 indicate the strong presence, moderate presence, weak presence, and absence of PU signal, respectively. Individual cluster generates a distinct binary data pattern, and these patterns are logically OR-ed for the first three clusters to obtain P_d and P_{fa}.

$$c_1 = \begin{cases} 1 & i, a^* = i, 1 \\ 0 & otherwise \end{cases}, c_2 = \begin{cases} 1 & i, a^* = i, 2 \\ 0 & otherwise \end{cases}, \quad (28)$$
$$c_3 = \begin{cases} 1 & i, a^* = i, 3 \\ 0 & otherwise \end{cases}, c_4 = \begin{cases} 1 & i, a^* = i, 4 \\ 0 & otherwise \end{cases}$$

Since SUs are not collocated, they are clustered based on some proximity. DE is used to obtain the optimum cluster centers for minimizing ω_c and N values. The objective of the proposed work is to determine the optimal $\{\mathcal{V}_1^*, \mathcal{V}_2^*, \mathcal{V}_3^*, \mathcal{V}_4^*\}$ to maximize the P_D value under the constraint of P_{fa}. Hence, the objective function can be mathematically expressed as

$$\{V_i^*\} = \max_{V_i} P_D \ , \quad \forall i = 1, 2, 3, 4$$

$$s.t. \ P_{fa} \leq P_{fa}^{th} \tag{29}$$

If the particular set of cluster provides $P_{fa} \leq P_{fa}^{th}$, then the corresponding set of clusters is included in the search space $(\hat{V})$ of DE algorithm. Hence, this constraint on the cluster selection in the search space actually reduces the size of initial population which in turn decreases the DE's computational overhead. DE obtains the optimal cluster centers for which ω_c and N values are minimal under the constraints of P_D and P_{fa}. The optimal values of ω_c and N minimize the overall power requirement in CSS. In brief, the integration of PFCM and DE clubs the benefits of individuals. While PFCM clusters the energy values, DE optimizes the cluster centers through the value of ω_c and N that in turn lead to energy minimization in CSS. Thus, formation of clusters for the energy values by PFCM under the constraints of detection probability reduces the search space of DE, which could have been extremely high without the use of PFCM.

4.1 DE Algorithm Based on PFCM Clustering

DE algorithm is a vector population-based meta-heuristics optimization method where population contains Np^* individuals. In the present work, DE is applied, instead of analytical approach, to obtain the global optimal set of cluster centers to improve the reliability in CSS at different SNR values. DE is applied in Eq. (18) to obtain the minimum ω_c and N values while satisfying the $P_{fa} \leq P_{fa}^{th}$ and $P_D \geq P_D^{th}$ constraints. Hence, the mathematical optimization problem for DE is formed as

$$\underset{\omega_c, N}{minimize} \ E_{PA} = \sum_{i=1}^{K} Power_i N(t_{sd})$$

$$s.t. \ P_{fa} \leq P_{fa}^{th},$$

$$P_D \geq P_D^{th} \tag{30}$$

The sequential steps of DE algorithm are mentioned in Algorithm 1.

4.1.1 DE Mutation Strategy

The solution in DE optimization depends to a certain extent on the choice of appropriate mutation strategy among the many existing ones. The widely used DE mutation strategies are $DE/Rand/1$, $DE/Rand/2$, $DE/Best/1$, $DE/Best/2$, and $DE/rand - best/1$ [45]. The extensive studies found that $DE/Rand/1$ and $DE/Best/2$ are the most effective among the other mutation strategies [46]. These

---

Input : A finite set of population Np^* individuals $\hat{\mathcal{V}} = \{\mathcal{V}_{i,1}, \mathcal{V}_{i,2}, \mathcal{V}_{i,3}, \mathcal{V}_{i,4}\}$ is determined to satisfy
$P_{fa} \leq P_{fa}^{th}$, where $i \in \{1, Np^*\}$.
The set of cluster centers $\{\mathcal{V}_1^*, \mathcal{V}_2^*, \mathcal{V}_3^*, \mathcal{V}_4^*\}$ is obtained by the DE algorithm. Typically the generation count C is fixed to 0 at the beginning of the evaluation process.
Output: Optimal $\{\mathcal{V}_1^*, \mathcal{V}_2^*, \mathcal{V}_3^*, \mathcal{V}_4^*\}$ are obtained at the final stage.
The sequential steps of DE algorithm are as follows:

---

begin
  repeat
    Step I: Mutation: Initiate a mutated vector $\overrightarrow{A}_i c$ corresponding to target vector $\overrightarrow{\mathcal{V}}_i, c \; \forall i$.
    for $i = 1$ *to* Np^* **do**
      for $j = 1$ *to* 4 **do**

$$A_{i,j}c = \mathcal{V}_{r_1}, j, c + F(\mathcal{V}_{r_2}, j, c - \mathcal{V}_{r_3}, j, c)$$

        where r_1, r_2, and r_3 are randomly chosen vectors from $\mathcal{V}$ and scaling factor F is =0.5, and $F \in [0, 1]$
      end
    end
    Step II: Crossover: For each target vector $\overrightarrow{\mathcal{V}}_i, c$, generates a trial vector $\overrightarrow{B}_i, c$ when a randomly assigned value $[0, 1] < (C_r)$.
    for $i = 1$ *to* Np^* **do**
      for $j = 1$ *to* 4 **do**

$$\overrightarrow{B}_{i,j}, c = \begin{cases} \overrightarrow{A}_{i,j}, c, & \text{if } rand_j(0,1) < C_r) \\ \overrightarrow{\mathcal{V}}_{i,j}, c, & \text{otherwise} \end{cases}$$

      end
    end
    Step III: Selection: Estimate the trial vector.
    for $i = 1$ *to* Np^* **do**

$$\overrightarrow{\mathcal{V}}_i, (c+1) = \begin{cases} \overrightarrow{B}_i, c & \text{if } P_D(\overrightarrow{B}_i, c) \geq P_D(\overrightarrow{\mathcal{V}}_i, c) \\ \overrightarrow{\mathcal{V}}_i, c, & \text{otherwise} \end{cases}$$

    end
    Increment the count C=C+1
  until *maximum iteration count is reached*;
end

Algorithm 1: DE algorithm based PFCM clustering

two mutation strategies are mathematically expressed as

$$Rand/1 \rightarrow A_{i,j}c = \mathcal{V}_{r_1}, j, c + F(\mathcal{V}_{r_2}, j, c - \mathcal{V}_{r_3}, j, c) \qquad (31)$$

$$Best/2 \rightarrow A_{i,j}c = \mathcal{V}_{r_1 best} + F(\mathcal{V}_{r_2}, j, c - \mathcal{V}_{r_3}, j, c + \mathcal{V}_{r_4}, j, c - \mathcal{V}_{r_5}, j, c) \qquad (32)$$

However, $DE/Best/2$ provides a faster convergence speed and it performs well on the uni-modal problems, but extensive exploitation of this strategy increases the probability of obtaining a local optimal solution [46]. On the other hand, the $DE/Rand/1$ adopts an over exploration strategy among the best chromosomes and provides the global optimal solution but reduces the convergence speed [46, 47]. Hence, $DE/Rand/1$ is applied in the proposed work under the maximum number of generation count (C) for obtaining the global optimal solution. The present work also reports the performance analysis between $DE/Rand/1$ and $DE/Best/2$ strategies in the numerical section for the fixed set of simulation parameters.

4.1.2 DE Control Parameters

There are three control parameters, namely initial population (Np^*), crossover rate (C_r), and scaling factor (F), mostly influence the convergence probability and the execution speed of the DE algorithm [48, 49].

(i) Initial Population: Population initialization affects the convergence speed and provides the better solution [36]. The DE population includes Np^* vectors. It is recommended by Storn and Price that the population size must be 10 times or higher than the dimensions of the input space [50]. The large number of population explores the all aspects of the search space for low-dimensional problems. However, due to the time constraints the population size must be restricted. After several experimental analysis, it is found that for $Np^* = 50$, the optimization function obtains the optimal solution for the present problem.

(ii) Crossover Rate: DE employs the C_r to generate a trial vector between the target and the mutant vectors to expand the current population. The typical binomial crossover is performed on each of the vectors [51]. In general, C_r is a user-specified constant within the range of [0, 1]. It is observed that for small C_r values the iteration count for saturating fitness value is increased. The selection of an efficient C_r value is totally based on the nature of the problem. It is observed that $C_r \in [0.9, 1]$ is appropriate for the non-separable objective functions while $C_r \in [0, 0.2]$ is effective for the separable functions [52]. After several experimental observations, the C_r value is fixed at 0.5 for the proposed work.

(iii) Scale Factor: The scale factor $(F \in [0, 1])$ amplifies the differences between the two vectors. In typical DE scheme, the F value remains constant during the program execution. The premature convergence takes place due to the small values of F, while the very high values of F delay the overall searching process [49]. The F value is widely varied along with the patterns of Np^* in different application problems [35, 48]. According to the [50], the F value is fixed at 0.5 in the proposed work while $Np^* > 10$.

5 Numerical Results and Analysis

Performances of the proposed work are substantially evaluated and reported in this section. It is observed that in similar framework, i.e., at low SNR, PFCM-based DE strategy offers better PU detection accuracy and reduces the average energy consumption over the existing techniques [25–28]. Due to the dynamic nature of wireless channels, a large number $\sim$10,000 times of independent iteration are performed to average a single outcome. Few parameters' values keep unchanged throughout the different experiments, such as $\mathcal{P}(\mathcal{H}_1) = 0.30$, $\mathcal{P}(\mathcal{H}_0) = 0.70$, $\alpha = 3$, $P_{noise} = 0$ dBW, mean $d_{aK} = 1.5$ m, and mean $d_{bK} = 2$ m.

5.1 PFCM Clustering in CSS

Figure 2a plots the receiver operating characteristics (ROC) curve of P_D versus P_{fa} at $P_{pow} = -10$ dBW, $K = 10$, and $N = 200$. It is noted that PFCM increases P_D values at $P_{fa} = 0.15$ by 9.511%, 16.14%, and 54.39% compared to [25, 28] and [27], respectively.

Figure 2b graphically represents P_D versus P_{fa} at $P_{pow} = -20$ dBW when $N = 500$ and $K = 12$. From figure, it is observed that proposed scheme improves the P_D values by $\sim 12.53\%$, $\sim 20.73\%$, and $\sim 44.42\%$ over the other techniques in [25, 28] and [27], respectively, at $P_{fa} = 0.30$.

Figure 3a demonstrates the change in P_D values with variation in K, while P_{pow} and N are 250 and -10 dBW, respectively. It is observed that PFCM technique improves P_D by $\sim 24.92\%$ when the number of SUs is also increased from 5 to 9.

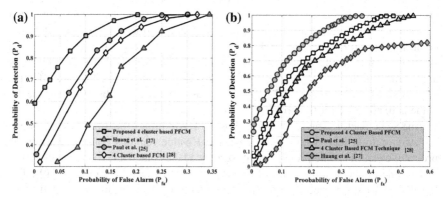

Fig. 2 **a** Receiver operating characteristics (ROC) curve for P_D versus P_{fa}, at $P_{pow} = -10$ dBW, number of SUs (K) = 10 and $N = 200$ **b** ROC plot for P_D versus P_{fa} at $P_{pow} = -20$ dBW, number of SUs (K) = 12, and $N = 500$

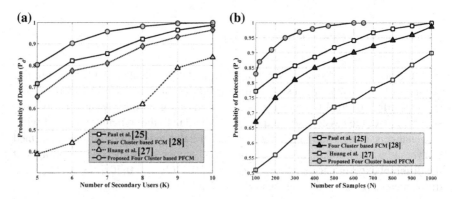

Fig. 3 **a** P_D versus K at $N = 250$ and $P_{pow} = -10$ dBW **b** plot P_D versus N at $K = 8$ and $P_{pow} = -10$ dBW

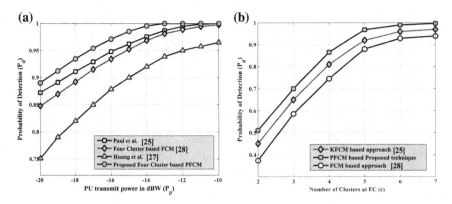

Fig. 4 **a** Plot of P_D versus (P_{pow}) at $N = 500$ and $K = 12$ **b** plot of P_D versus count of clusters c at $N = 250$, $P_{pow} = -15$ dBW, and $K = 10$

From the figure, it is noted that PFCM outperforms P_D values by $\sim 3.59\%$, $\sim 7.14\%$, and $\sim 24.96\%$ over the existing schemes in [25, 28] and [27], respectively, at $K = 9$.

Figure 3b illustrates that PFCM improves P_D by $\sim 17.12\%$ when N is increased from 100–400. PFCM also enhances P_D values by $\sim 6.26\%$, $\sim 10.98\%$, and $\sim 23.45\%$ over KFCM [25], FCM [28] and [27], respectively, at $N = 600$.

Figure 4a demonstrates the changes in P_D along with (P_{pow}) values when N and K are fixed. PFCM increases P_D by $\sim 11.23\%$ when P_{pow} value is increased from -20 to -14 dBW. PFCM enhances P_D values by $\sim 2.60\%$, $\sim 4.56\%$, and $\sim 9.44\%$ compared to [25, 28] and [27], respectively, at $P_{pow} = -15$ dBW.

Figure 4b represents that P_D is improved along with the increment in the number of clusters. PFCM enhances the P_D by $\sim 76.03\%$ when clusters are increased from 2 to 4 at $K = 10$, $P_{pow} = -15$ dBW, and $N = 250$. The proposed scheme also improves P_D values by $\sim 7.47\%$ and $\sim 12.98\%$ over [25] and [28], respectively, at $c = 4$.

5.2 PFCM and DE for Energy Efficient Sensing

This subsection reports the power consumption by different algorithms during CSS. The simulation parameters are fixed at $d_{aK} = 1$ m, $d_{bK} = 2$ m, $P_{noise} = 0$ dBW, $N = [25, \ldots, 1000]$, $P_{pow} = 0$ dBW, $\omega_c = \{2, \ldots, 100\}$, $\alpha = 4$, $\mathcal{P}(\mathcal{H}_0) = 0.70$, $\mathcal{P}(\mathcal{H}_1) = 0.30$, $T_{sd} = 1$ ms, and $K = 10$. The constraints $P_D \geq 0.90$ and $P_{fa} \lesssim 0.05$ are imposed on all cases.

Figure 5a illustrates the changes in E_{cons} along with the variations in ω_c. It is observed for all the techniques that the E_{cons} is decreased initially with an increment in ω_c value. This drop in E_{cons} is obtained due to an immediate reduction in N value until ω_c achieves its minimal value and after that E_{cons} is consistently increased. From

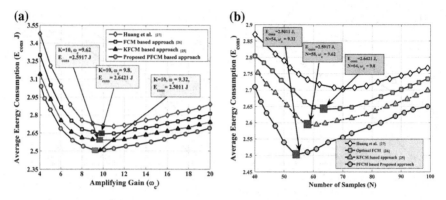

Fig. 5 **a** E_{cons} versus ω_c at $P_D \geq 0.90$, $P_{fa} \leq 0.05$, $t_{sd} = 1$ ms **b** E_{cons} versus (N) at $K = 10$, $P_D \geq 0.90$, $P_{fa} \leq 0.05$, $t_{sd} = 1$ ms

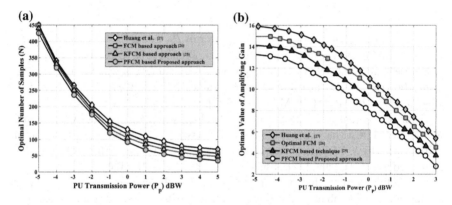

Fig. 6 **a** N versus P_{pow} at $K = 10$ **b** ω_c versus P_{pow} at $K = 10$

the figure, it is noted that optimal-FCM- [26] and KFCM [25]-based strategies obtain the optimal ω_c at 9.8 and 9.62 while E_{cons} are 2.6421 J and 2.5917 J, respectively. It is worth mentioning that PFCM minimizes E_{cons} by $\sim 3.97\%$ and $\sim 5.34\%$ compared to [25] and [26], respectively.

Figure 5b shows that the proposed technique obtains an optimal $N = 54$ while [25] and [26] find optimal N at 58 and 64, respectively. Therefore, PFCM offers faster sensing by $\sim 6.89\%$ and $\sim 15.56\%$ over [25] and [26], respectively.

Figure 6a, b represents the changes in N and ω_c, respectively, along with the variation in P_{pow}. It is observed that increment in P_{pow} value decreases the optimal N and ω_c values. PFCM-based DE technique offers more gain in ω_c by $\sim 9.89\%$, $\sim 19.60\%$, $\sim 29.91\%$ and in N by $\sim 12.38\%$, $\sim 20.68\%$, $\sim 29.23\%$ over [25], [26] and [27], respectively, at $P_{pow} = 0$ dBW.

Figure 7 graphically demonstrates the change in the total number of PU samples with the amount of required amplifying gain (ω_c) while SNR and P_D threshold are

Fig. 7 Number of samples
(N) versus optimal value of
amplifying gain (ω_c)

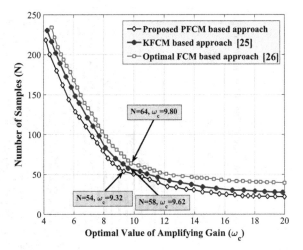

Fig. 8 Optimal average
energy consumption versus
probability of detection

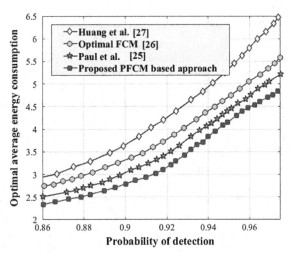

fixed at 0 dBW and 0.95, respectively. It is clearly observed that the increment in the
(ω_c) values reduces the number of PU samples to meet the predefined P_D constraint.
It is noted that the proposed work minimizes the required number of samples by
$\sim 9.89\%$ and $\sim 23.74\%$ when compared to the existing works in [25] and [26],
respectively, at ω_c=10.

Figure 8 illustrates the variation in the optimal average energy consumption
against the PU signal detection probability. From the graphical results, it is observed
that average energy consumption is increased with the increase in the values of P_D.
In the similar framework, the proposed work outperforms the existing techniques in
[25–27] by $\sim 7.90\%$, $\sim 17.07\%$, and $\sim 31.42\%$, respectively, at $P_D = 0.90$.

According to the central limit theorem (CLT) for the large number of sample
values, ($E_{Max_{final}}$) at FC follows the Gaussian distribution with different mean and

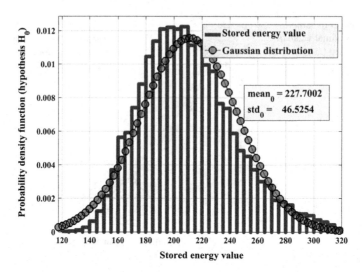

Fig. 9 Probability density function (hypothesis H_0) versus stored energy values ($E_{Max_{final}}$)

variance for the respective hypotheses. Figures 9 and 10 illustrate the probability density function (pdf) versus the energy values stored in $E_{Max_{final}}$ for the hypotheses H_0 and H_1, respectively. It is worth mentioning that the pdf of the $E_{Max_{final}}$ follows Gaussian distribution. It is noted from Fig. 9 that the statistical mean and standard deviation (SD) for the H_0 hypothesis are 227.7002 and 46.5254, respectively. Similarly from Fig. 10 for H_1 hypothesis, the mean and SD values are 229.0714 and 46.7694, respectively.

Figure 11 demonstrates the ROC plot for the probability of PU detection versus its P_{fa}. P_D is plotted for different values of P_{fa}, while the reporting channel is considered to be followed the Rayleigh fading. It is observed that for the higher values of Np^*, C_r, and F, the P_D value is increased over the fixed P_{fa}. In this case, the P_{pow} is 0 dBW while the values of N and K are fixed at 150 and 12, respectively. It may be observed that $P_D\{N_p^* = 20, C_r = 0.3, F = 0.3\} < P_D\{N_p = 40, C_r = 0.4, F = 0.4\} < P_D\{N_p = 50, C_r = 0.5, F = 0.5\} < P_D\{N_p = 65, C_r = 0.55, F = 0.55\}$ for SU-FC reporting scenario. It is worth mentioning that increasing the values of N_p^* increases the P_D values but it significantly increases the searching time for the optimal solution. In the present work, it is noted that for $N_p^* = 50$ and $N_p^* = 65$ the system obtains almost similar optimal solution. Hence, the N_p^* value is fixed at 50.

Figure 12 demonstrates the ROC plot for the maximum P_D versus its P_{fa} for the two different mutation strategies, namely $/DE/Rand/1$ and $/DE/Best/2$ when $N_p^* = 500$. The large size of N_p^* space helps to distinguish the performance difference between the above-mentioned mutation strategies. In this case, it is observed that both mutation strategies perform well and provide similar optimal values but in few cases $/DE/Best/2$ is stuck to the local optimal solution while $/DE/Rand/1$ provides the global solution. From Fig. 12, it is noted that $/DE/Best/2$ provides P_D value

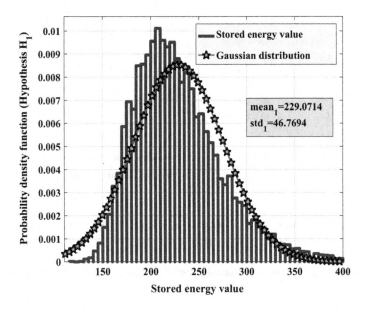

Fig. 10 Probability density function (hypothesis H_1) versus stored energy values ($E_{Max_{final}}$)

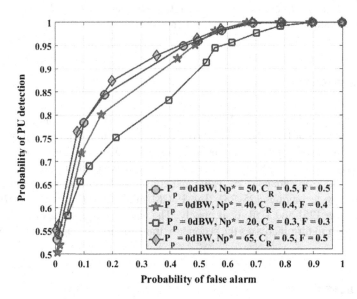

Fig. 11 Probability of PU detection versus probability of false alarm (for different DE parameter values)

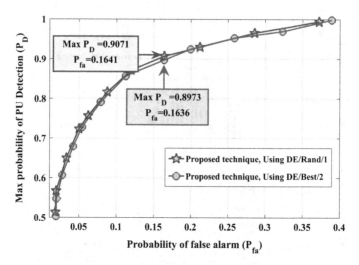

Fig. 12 Probability of PU detection versus probability of false alarm (for different DE mutation strategies, when $N_p^* = 500$)

0.8973 when respective P_{fa} is 0.1636. On the other hand, $/DE/Rand/1$ provides the optimal value of P_D that is 0.9071 when respective P_{fa} is 0.1641.

6 Conclusions

The present work exploits PFCM clustering algorithm in CSS technique. The performance of CSS at low SNR is improved compared to the techniques proposed in [25–28]. PFCM offers higher gain in the P_D by $\sim 12.53\%$, $\sim 20.73\%$, and $\sim 44.42\%$ at -20 dBW over [25, 28] and [27], respectively. It is noted that in every aspects PFCM provides an enhanced P_D value and consumes less energy compared to the other existing techniques [25–28]. The PFCM minimizes E_{cons} by $\sim 3.97\%$ and $\sim 5.34\%$ over [25] and [26], respectively. The faster sensing is also achieved by PFCM which in turn increases the secondary transmission throughput in CRN.

References

1. Federal Communications Commission: Spectrum policy task force, Rep. ET Docket no. 02-135 (2002)
2. OFCOM: Digital Dividend Review. A statement on our approach towards awarding the digital dividend (2007)
3. Mitola, J.: Cognitive radio: an integrated agent architecture for software defined radio. Ph.D. dissertation, Computer Communication System Laboratory, Department of Teleinformatics,

Royal Institute of Technology (KTH), Stockholm, Sweden, May 2000

4. Haykin, S., Setoodeh, P.: Cognitive radio networks: the spectrum supply chain paradigm. IEEE Trans. Cognit. Commun. Netw. **1**(1), 3–28 (2015)

5. Banerjee, A., Paul, A., Maity, S.P.: Joint power allocation and route selection for outage minimization in multihop cognitive radio networks with energy harvesting. IEEE Trans. Cognit. Commun. Netw. **4**(1), 82–92 (2018)

6. Paul, A., Maity, S.P.: On outage minimization in cognitive radio networks through routing and power control. Wirel. Pers. Commun. **98**(1), 251–269 (2018)

7. Bhatti, D.M.S., Nam, H.: Spatial correlation based analysis of soft combination and user selection algorithm for cooperative spectrum sensing. IET Commun. **11**(1), 39–44 (2017)

8. Banerjee, A., Maity, S.P.: On optimal sample checkpoint for energy efficient cooperative spectrum sensing. Digit. Signal Process. **74**, 56–71 (2018)

9. Sobron, I., Diniz, P., Martins, W., Velez, M.: Energy detection technique for adaptive spectrum sensing. IEEE Trans. Commun. **63**(3), 617–627 (2015)

10. Shen, J., Jiang, T., Liu, S., Zhang, Z.: Maximum channel throughput via cooperative spectrum sensing in cognitive radio networks. IEEE Trans. Wirel. Commun. **8**(10), 5166–5175 (2009)

11. Mingchuan, Y., Yuan, L., Xiaofeng, L., Wenyan, T.: Cyclostationary feature detection based spectrum sensing algorithm under complicated electromagnetic environment in cognitive radio networks. China Commun. **12**(9), 35–44 (2015)

12. Xinzhi, Z., Feifei, G., Rong, C., Tao, J.: Matched filter based spectrum sensing when primary user has multiple power levels. China Commun. **12**(2), 21–31 (2015)

13. Zeng, Y., Liang, Y.C.: Eigenvalue-based spectrum sensing algorithms for cognitive radio. IEEE Trans. Commun. **57**(6), 1784–1793 (2009)

14. Zhang, Y., Zhang, Q., Wu, S.: Entropy-based robust spectrum sensing in cognitive radio. IET Commun. **4**(4), 428–436 (2010)

15. Xu, Y.L., Zhang, H.S., Han, Z.H.: The performance analysis of spectrum sensing algorithms based on wavelet edge detection. In: Proceeding of 5th International Conference on Wireless Communications, Networking and Mobile Computing (WiCom), pp. 1–4 (2009)

16. Sedighi, S., Taherpour, A., Monfared, S.: Bayesian generalised likelihood ratio test-based multiple antenna spectrum sensing for cognitive radios. IET Commun. **7**(18), 2151–2165 (2013)

17. Sun, W., Huang, Z., Wang, F., Wang, X.: Compressive wideband spectrum sensing based on single channel. Electron. Lett. **51**(9), 693–695 (2015)

18. Jaglan, R.R., Sarowa, S., Mustafa, R., Agrawal, S., Kumar, N.: Comparative study of single-user spectrum sensing techniques in cognitive radio networks. Procedia Comput. Sci. **58**, 121–128 (2015)

19. Yucek, T., Arslan, H.: A survey of spectrum sensing algorithms for cognitive radio applications. IEEE Commun. Surv. Tutor. **11**(1), 116–130 (2009)

20. Bhargavi, D., Murthy, C.: Performance comparison of energy, matched-filter and cyclostationarity-based spectrum sensing. In: Proceeding IEEE 11th International Workshop on Signal Processing Advances in Wireless Communications (SPAWC), pp.1–5 (2010)

21. So, J.: Energy-efficient cooperative spectrum sensing with a logical multi-bit combination rule. IEEE Commun. Lett. **20**(12), 2538–2541 (2016)

22. Awin, F.A., Abdel-Raheem, E., Ahmadi, M.: Designing an optimal energy efficient cluster-based spectrum sensing for cognitive radio networks. IEEE Commun. Lett. **20**(9), 1884–1887 (2016)

23. Cichoń, K., Kliks, A., Bogucka, H.: Energy-efficient cooperative spectrum sensing: A survey. IEEE Commun. Surv. Tutor. **18**(3), 1861–1886

24. Jiao, Y., Yin, P., Joe, I.: Clustering scheme for cooperative spectrum sensing in cognitive radio networks. IET Commun. **10**(13), 1590–1595 (2016)

25. Paul, A., Maity, S.P.: Kernel fuzzy c-means clustering on energy detection based cooperative spectrum sensing. Digital Commun. Netw. **2**(4), 196–205 (2016)

26. Maity, S.P., Chatterjee, S., Acharya, T.: On optimal fuzzy c-means clustering for energy efficient cooperative spectrum sensing in cognitive radio networks. Digital Signal Process. **49**(C), 104–115

27. Huang, S., Chen, H., Zhang, Y., Zhao, F.: Energy-efficient cooperative spectrum sensing with amplify-and-forward relaying. IEEE Commun. Lett. **16**(4), 450–453 (2012)
28. Chatterjee, S., Banerjee, A., Acharya, T., Maity, S.P.: Fuzzy c-means clustering in energy detection for cooperative spectrum sensing in cognitive radio system. Proc. Mult. Access Commun. **8715**, 84–95 (2014)
29. Graves, D., Pedrycz, W.: Kernel-based fuzzy clustering and fuzzy clustering : A comparative experimental study. Fuzzy Sets Syst. **161**(4), 522–543 (2010)
30. Shawe-Taylor, J., Cristianini, N.: Kernel Methods for Pattern Analysis. Cambridge University Press, New York, NY, USA (2004)
31. Zhao, X., Zhang, S.: In: An Improved KFCM Algorithm Based on Artificial Bee Colony, pp. 190–198. Springer, Berlin Heidelberg, Berlin, Heidelberg (2011)
32. Pal, N.R., Pal, K., Keller, J.M., Bezdek, J.C.: A possibilistic fuzzy c-means clustering algorithm. IEEE Transactions on Fuzzy Systems **13**(4), 517–530 (2005)
33. Shang, R., Tian, P., Wen, A., Liu, W., Jiao, L.: An intuitionistic fuzzy possibilistic c-means clustering based on genetic algorithm. In: Proceedings of IEEE Congress on Evolutionary Computation (CEC), July 2016, pp. 941–947
34. Paul, A., Maity, S.P.: On energy efficient cooperative spectrum sensing using possibilistic fuzzy c-means clustering. In: Intelligence, Computational (ed.) Communications, and Business Analytics, pp. 382–396. Springer Singapore, Singapore (2017)
35. Gao, W., Yen, G.G., Liu, S.: A cluster-based differential evolution with self-adaptive strategy for multimodal optimization. IEEE Trans. Cybern. **44**(8), 1314–1327 (2014)
36. Wang, J., Zhang, W., Zhang, J.: Cooperative differential evolution with multiple populations for multiobjective optimization. IEEE Trans. Cybern. **46**(12), 2848–2861 (2016)
37. Saha, A., Konar, A., Rakshit, P., Ralescu, A.L., Nagar, A.K.: Olfaction recognition by eeg analysis using differential evolution induced hopfield neural net. In: Proceedings of International Joint Conference on Neural Networks (IJCNN), August 2013, pp. 1–8
38. Bhattacharyya, S., Rakshiti, P., Konar, A., Tibarewala, D.N., Das, S., Nagar, A.K.: Differential evolution with temporal difference q-learning based feature selection for motor imagery eeg data. In: Proceedings of IEEE Symposium on Computational Intelligence, Cognitive Algorithms, Mind, and Brain (CCMB), April 2013, pp. 138–145
39. Iliya, S., Goodyer, E., Shell, J., Gongora, M., Gow, J.: Optimized neural network using differential evolutionary and swarm intelligence optimization algorithms for rf power prediction in cognitive radio network: A comparative study. In: Proceedings of IEEE 6th International Conference on Adaptive Science Technology (ICAST), October 2014, pp. 1–7
40. Anumandla, K.K., Peesapati, R., Sabat, S.L., Udgata, S.K., Abraham, A.: Field programmable gate arrays-based differential evolution coprocessor: a case study of spectrum allocation in cognitive radio network. IET Comput. Digital Tech. **7**(5), 221–234 (2013)
41. Anumandla, K.K., Akella, B., Sabat, S.L., Udgata, S.K.: Spectrum allocation in cognitive radio networks using multi-objective differential evolution algorithm. In: Proceedings of International Conference on Signal Processing and Integrated Networks (SPIN), February 2015, pp. 264–269
42. Lina, C.: Power control algorithm for cognitive radio based on differential evolution. In: Proceedings of International Conference on Computer Application and System Modeling (ICCASM 2010), Vol. 7, October 2010, V7-474–V7-478
43. Zhang, X., Zhang, X.: Population-adaptive differential evolution-based power allocation algorithm for cognitive radio networks. EURASIP J. Wirel. Commun. Netw. **2016**(1), 219 (2016)
44. Almeida, R.J., Kaymak, U., Sousa, J.M.C.: Fuzzy rule extraction from typicality and membership partitions. In: Proceedings of IEEE International Conference on Fuzzy Systems (IEEE World Congress on Computational Intelligence, June 2008, pp. 1964–1970
45. Hu, C., Yan, X.: A hybrid differential evolution algorithm integrated with an ant system and its application. Comput. Math. Appl. **62**(1), 32–43 (2011)
46. Nobakhti, A., Wang, H.: A simple self-adaptive differential evolution algorithm with application on the alstom gasifier. Appl. Soft Comput. **8**(1), 350–370 (2008)

47. Epitropakis, M.G., Tasoulis, D.K., Pavlidis, N.G., Plagianakos, V.P., Vrahatis, M.N.: Enhancing differential evolution utilizing proximity-based mutation operators. IEEE Trans. Evol. Comput. **15**(1), 99–119 (2011)
48. Sarker, R.A., Elsayed, S.M., Ray, T.: Differential evolution with dynamic parameters selection for optimization problems. IEEE Trans. Evol. Comput. **18**(5), 689–707 (2014)
49. Li, X., Yin, M.: Modified differential evolution with self-adaptive parameters method. J. Comb. Optim. **31**(2), 546–576 (2016)
50. Storn, R., Price, K.: Differential evolution–a simple and efficient heuristic for global optimization over continuous spaces. J. Global Optim. **11**(4), 341–359 (1997)
51. Gong, W., Cai, Z., Wang, Y.: Repairing the crossover rate in adaptive differential evolution. Appl. Soft Comput. **15**, 149–168 (2014)
52. Mohamed, A.W., Sabry, H.Z., Abd-Elaziz, T.: Real parameter optimization by an effective differential evolution algorithm. Egypt. Inf. J. **14**(1), 37–53 (2013)

Feature Selection for Handwritten Word Recognition Using Memetic Algorithm

Manosij Ghosh, Samir Malakar, Showmik Bhowmik, Ram Sarkar and Mita Nasipuri

Abstract Nowadays, feature selection is considered as a de facto standard in the field of pattern recognition where high-dimensional feature attributes are used. The main purpose of any feature selection algorithm is to reduce the dimensionality of the input feature vector while improving the classification ability. Here, a Memetic Algorithm (MA)-based wrapper–filter feature selection method is applied for the recognition of handwritten word images in segmentation-free approach. In this context, two state-of-the-art feature vectors describing texture and shape of the word images, respectively, are considered for feature dimension reduction. Experimentation is conducted on handwritten Bangla word samples comprising 50 popular city names of West Bengal, a state of India. Final results confirm that for the said recognition problem, subset of features selected by MA produces increased recognition accuracy than the individual feature vector or their combination when applied entirely.

Keywords Feature selection · Memetic algorithm · Wrapper–filter method
Handwritten word recognition · Bangla script · City name recognition

M. Ghosh (✉) · S. Bhowmik · R. Sarkar · M. Nasipuri
Department of Computer Science and Engineering, Jadavpur University, Kolkata, India
e-mail: manosij1996@gmail.com

S. Bhowmik
e-mail: showmik.cse@gmail.com

R. Sarkar
e-mail: ramsarkar@gmail.com

M. Nasipuri
e-mail: mitanasipuri@gmail.com

S. Malakar
Department of Computer Science, Asutosh College, Kolkata, India
e-mail: malakarsamir@gmail.com

© Springer Nature Singapore Pte Ltd. 2019   103
J. K. Mandal et al. (eds.), *Advances in Intelligent Computing*,
Studies in Computational Intelligence 687,
https://doi.org/10.1007/978-981-10-8974-9_6

1   Introduction

One of the major prerequisites in solving any pattern recognition problem is to design a befitting feature vector that can uniquely represent the patterns in the feature space. Therefore, a large number of researchers over the years have been devoted to fix up various pattern classification problems by introducing several new/modified feature vectors based on shape, texture or topology of the patterns. As a result, designing more and more features has become a common trend. But, increased dimension of feature vectors may not always provide better outcome, as generation of huge number of features does not ensure the orthogonal property of features in the feature space. The key reason for this is that feature values may have redundant and irrelevant information. Another limitation of the high-dimensional feature vector is that time required to build a recognition module for classifying a data increases since this time is directly proportional to the feature dimension under consideration.

In addition to these, combining two or more good features may not yield better result unless the features to be combined have the ability to provide some complimentary information about the patterns to be classified. However, it is also true that extracting features from patterns using different approaches provide some new information about the patterns. But identifying such informative feature set is always a complex research problem. Here comes the usefulness of the feature selection algorithms. Feature selection/reduction refers to the study of algorithms for reducing the number of dimensions of data representing the pattern classes. The main purpose of this is to identify fewer features to represent the patterns and reduce the computational cost, without weakening discriminative capability of the same. Subsequently, feature reduction can be proven to be beneficial in the later stages of the learning algorithms, such as dodging over-fitting, resisting noise and augmenting prediction performance.

However, the problem of feature reduction can be stated as a problem of selecting the best feature subset from the search space. Selecting a subset from a set is an NP-hard problem; i.e., there exist $2^n - 1$ number of possible subsets for an n-element feature set. Such presence of exponential possible subsets in the solution space drives the researchers towards using some optimizing techniques. In this work, an MA-based feature selection method is used to find out the optimal number of features to be used for recognizing handwritten Bangla words. This paper is an extended version of the work reported in [1].

1.1   Feature Selection Approaches

Feature selection approaches, as found in the literature, are of three types, namely filter [2, 3], wrapper [4–8] and embedded [9–13]. The algorithms that follow first approach, in general, select a subset of features in unsupervised means. Statistical measures like mean, standard deviation, correlation coefficient and statistical tests [14] like analysis of variance (ANOVA), chi-square are used to take decision whether to include a feature under consideration into the optimal set or not. Methods of this approach consider only the feature values and thereby generate rank of the features based on dissimilarity score. The ranks as well as the scores are used to obtain final feature set. Some of the well-known algorithms that follow filter approach are ReliefF [15], mutual information [16], symmetric uncertainty [17], etc.

On the contrary, algorithms that belong to wrapper category apply some supervised learning mechanism (to get the score of the objective function) in order to decide whether to keep some features into final feature set or not. Incorporation of learning tool eventually increases the feature selection time. But, this approach generally comes up with better classification accuracy. The prime reason for this is the use of performance of the supervised learning mechanism in the objective function. Some of the popularly used wrapper feature selection techniques are genetic algorithm (GA) [4–6], particle swarm optimization (PSO) [7], adaptive neuro-fuzzy inference system [8], branch and bound (BB) [18], ant colony optimization (ACO) [19] etc.

From the above discussion, it is clear that the filter methods provide ranking of a feature in a data-independent way, while the wrapper methods have objective function (a learning algorithm) dependency in selecting the optimal feature set. Apart from these two approaches, researchers have introduced an alternative feature selection approach called embedded or filter–wrapper approach where the benefits of both wrapper and filter methods are considered.

MA as a wrapper–filter feature selection approach that has been applied for several real-time applications like pattern classification problem [13, 20, 21], instruction detection [22], stopping of premature classification convergence [23], prototype selection [24] and prediction problem [25, 26]. University of California Irvine (UCI) machine learning repository [27], microarray dataset [28] data are used for the said work. In addition to these, studies reveal that different variants of MA has been used in many other applications such as solution to training of artificial neural networks, robotic motion planning, circuit design, medical expert systems, manpower scheduling, VLSI design. Considering the generalness of MA-based feature selection, here, in the present work, the same has been chosen as feature selection method and has been applied to recognize handwritten word images.

1.2 Handwritten Word Recognition

Handwritten word recognition (HWR), one of the important branches related to content analysis in handwritten documents, aims to convert the word images present in a document page into machine-encoded form. A number of research attempts [29–36] have been tried by the researchers to perform HWR. From methodological point of view, the approaches used to solve the problem of HWR can broadly be classified into two categories such as analytical [29, 37] and holistic [30–36]. The former category of approaches performs explicit segmentation of word samples to generate several structurally similar sub-parts, known as lexeme/character, sometimes using script-specific knowledge. Then recognition–reconstruction model is used to build the machine-encoded form of the words. In this type of model, the segmented characters/lexemes are first recognized using some classifier. Then, these recognized characters/lexemes are merged based on their positional information [29] to form the machine-encoded word. But, this approach, in general, faces problems related to segmentation when word images are cursive in nature. As a result, for unconstrained handwriting, these methods need some post-processing based on lexicon to produce expected outcome. Hence, analytical HWR methods become computationally expensive.

On the other hand, holistic method [30–37] considers each word of the lexicon as an individual pattern class and recognizes them in a segmentation-free way. But in this case, designing some effective features becomes more complex than its counterpart. The reason behind this is that there are large variations in structure and shape among multiple instances belonging to same word written by different writers. Even the variations are notable for individual writer. But these techniques mostly show better performance when applied on limited and pre-defined lexicon. Some real-life applications of these approaches are city name recognition for postal automation [36], legal amount recognition for automating banking system [37], medicine name searching in doctor's prescription [38] etc.

1.3 Feature Selection in Bangla HWR

Bangla is a widely used language with around 250 million speakers worldwide [39]. The script contains 11 vowels and 39 consonants as basic character. In addition to this, the presence of compound characters, modified shapes, numerals and special symbols makes the Bangla alphabet set large. All these taken together have made unconstrained Bangla HWR a more complex research problem. Irrespective of the approaches followed to perform HWR, only a few researches [35, 36] have paid attention on reducing the feature dimension to eliminate redundant or misleading features. Interestingly, these methods have mainly tried to minimize the feature dimension by reducing the image blocks using Gaussian down-sampling technique which leads to hypothetical deletion of some image regions. This fact exposes that, in general, research to find some solutions for the said problem is carried out by introducing transformed features of smaller dimension, which may lead to information loss. Even, different sets of good features have been combined in some works [32, 35] in order to generate better feature descriptor, but that may contain redundant and/or irrelevant features. Here arises the issue of feature dimension reduction for Bangla HWR.

Keeping these facts in mind, in this paper, an embedded feature selection technique for Bangla HWR is described, which is an extended version of the work reported in [1]. Here, the extension is carried out in terms of pattern classes, use of another feature vector which is modified version of statistical and contour-based feature (SCF) [40] and combining it with original feature vector (i.e. gradient-based feature).

2 Present Work

There are many notable feature selection methods of which the use of embedded or a wrapper–filter combination is usually found to be the more useful than its wrapper or filter counterparts. In this work, MA-based wrapper–filter feature selection is applied on an in-house dataset of handwritten words. The data collection and preparation is described in the next subsection followed by a description of the features extraction, and finally, the description of the MA is provided. The entire process of the present work is depicted in Fig. 1.

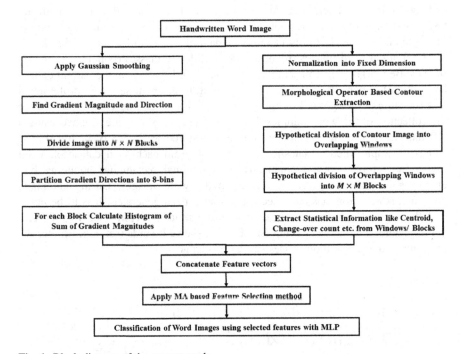

Fig. 1 Block diagram of the present work

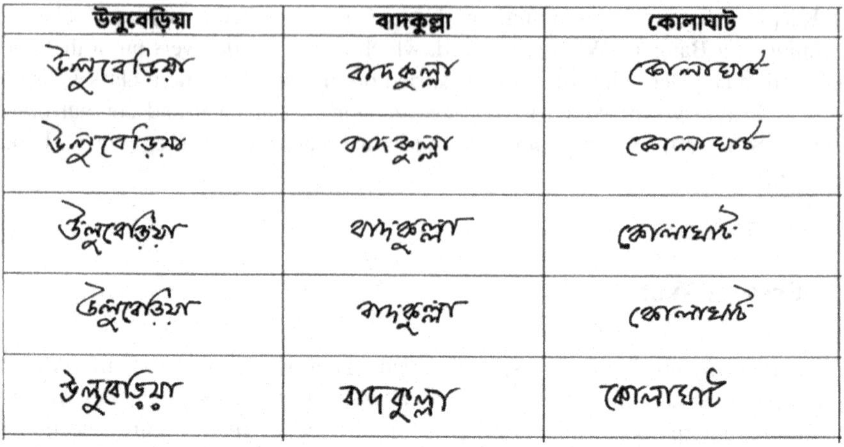

উলুবেড়িয়া	বাদকুল্লা	কোলাঘাট
উলুবেড়িয়া	বাদকুল্লা	কোলাঘাট
উলুবেড়িয়া	বাদকুল্লা	কোলাঘাট
উলুবেড়িয়া	বাদকুল্লা	কোলাঘাট
উলুবেড়িয়া	বাদকুল্লা	কোলাঘাট
উলুবেড়িয়া	বাদকুল্লা	কোলাঘাট

Fig. 2 Sample scanned datasheet with handwritten word images written in Bangla

2.1 Data Collection and Preparation

Data collection, in pattern recognition, is a weary but perhaps the most significant task. Because without appropriate data, any algorithm is converted to a purely theoretic endeavour. The present in-house dataset consists of a set of 7500 handwritten Bangla word images. It comprises 50 different classes (popular city names of West Bengal) having 150 sample variations. Word samples in the current dataset are written by more than 250 people belonging to different sex, age, professional and educational background. A pre-structured A4 size datasheet was designed for the collection of the word samples. Then, all these datasheets are scanned using a HP flatbed scanner in 300 dpi resolution and stored as 24 bit BMP file (Fig. 2 shows a sample filled-in datasheet). After that, from each such datasheet, word images are cropped programmatically with minimal rectangular bounding box enclosing all the data pixels belonging to the word images. During automatic cropping, some erroneous data (see Fig. 3) have been generated. It is to be noted that this dataset contains words with almost similar structure (see Fig. 4), city names having alternative spelling (see Fig. 5), skewed words (see Fig. 6), etc. All these make the recognition process a real challenge.

(a) **(b)** **(c)**

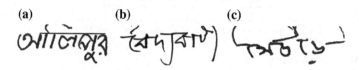

Fig. 3 Examples of errors (ascendants or descendants are partially cropped out) occur during automatic cropping of images from datasheet

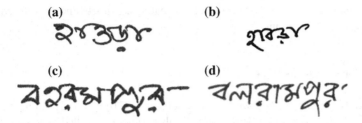

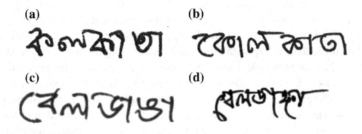

Fig. 4 Examples of structurally similar words (**a–b**) and (**c–b**)

Fig. 5 Instances of words having alternative spelling (**a** and **b**) and (**c** and **b**)

Fig. 6 Instances of skewed word images present in the current dataset

2.2 Feature Extraction

As stated earlier, the prime objective of the current work is to design a model (here, MA based) which can improve the performance of any pattern classification system by reducing the feature dimension using a feature selection method. A primary prerequisite for such work is to select proper features [41]. Therefore, as a case study, two recently developed features (i.e. gradient-based feature [30] and modified SCF [40]) that are designed for handwritten word recognition are considered here. Such choice of the feature vectors is made to get the complementary information about the pattern classes to be classified. First one uses gradient orientation information to represent a pattern in feature space, while the second one extracts some statistical information from contour of the shape of a pattern under consideration. This justifies the consideration of two different categories of feature vectors for combination. A brief description of the features, extracted here, is described in the following subsections.

2.2.1 Gradient-Based Features

Inspired by the concept reported in [42], authors of [30] had designed a grid-based gradient feature descriptor. This feature is also sometimes called as texture feature [43]. For this purpose, first, the word images are smoothed using Gaussian filter (see Fig. 7b). Then, gradient values in vertical and horizontal directions are calculated. Next, gradient magnitude (see Fig. 7c) and directions (see Fig. 7d) are computed. To extract the feature values, word images are divided into N × N hypothetical grids. Some of the word samples that are segmented hypothetically into 6 × 6 sub-images along with their gradient and magnitude values are shown in Fig. 8a–f. These figures reflect the importance of hypothetical image segmentation in terms of information available therein. In each segment, the pixels are classified into eight different categories based on their gradient orientation. Subsequently, eight feature values are generated by adding gradient magnitude value of the pixels of same category. Therefore, from each word image, a feature vector of length $8*N*N$ is generated.

2.2.2 Statistical and Contour Based Feature (SCF)

The SCF and moment-based features are introduced by Tamen et al. [40] to recognize Arabic handwritten words in a holistic way. In the present work, it is considered as a shape descriptor. It aims to extract local information of word images based on their contour information. It is to be noted that, here, few alterations in

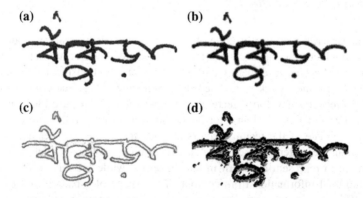

Fig. 7 Representation of different forms of pre-processed images generated during gradient-based feature extraction: **a** input word image, **b** image after applying Gaussian smoothing, **c** magnitude image **d** and image representing direction

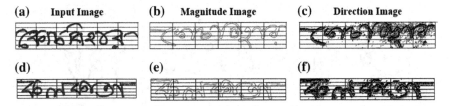

(a) **Input Image**   (b) **Magnitude Image**   (c) **Direction Image**

(d)   (e)   (f)

Fig. 8 Hypothetically segmented 6×6 sub-image structure of two city name images (**a** and **d**) and their gradient magnitude (**b** and **e**) orientations (**c** and **f**)

comparison with the actual implementation of SCF are made. The changes are made in terms of contour estimation, direction of sliding the window which is made in the context of Bangla script, choice of directions for estimating distance of the farthest contour pixel from centroid along with number of intersection points between contour and a line drawn in these directions. Some of the said parameter values are set experimentally. To extract the features, first, word images are normalized into fixed dimension (H_R, W_R) and then contour of normalized word image is estimated using morphological operator [44] (see Fig. 9). The local information extraction from this contour word image is described below.

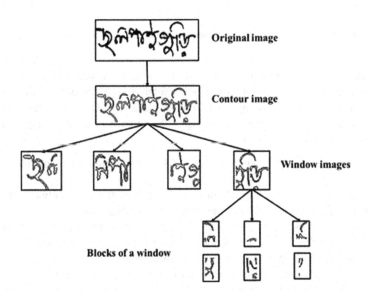

Fig. 9 Generation of sub-image structure from a word image (these sub-image structures are used to extract SCF values)

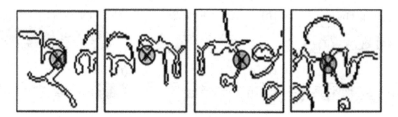

Fig. 10 Illustration of centroid positions inside overlapping sliding windows of a word image

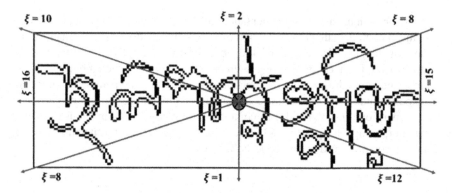

Fig. 11 Illustration of variations in ξ—values in 8-Freeman directions

To extract local information, at the outset, a window of width (W_{SW}) and height same as height of the contour image is considered. This window is slid over the contour image from left to right with OW_{SW} pixels overlap. An instance of generated window images is depicted in Fig. 9. From each of these sliding windows, the following features are extracted

- Centre of gravity (CG). The position of CG in each sliding window is shown in Fig. 10.
- Density of contour pixels which is computed as the ratio between the total number of contour pixels and the area of the sliding window.
- Number of intersection points (say, ξ) between the contour pixels and the lines drawn from the CG in d Freeman directions [45]. The variation in ξ values considering $d = 8$ in a word image is shown in Fig. 11.
- Distance of the farthest point in each of the d directions (see Fig. 12).
- For each of the sliding windows, density and the number of data to background pixels transition in both vertical and horizontal directions from $m \times n$ blocks (2×3 blocks in each window is shown in Fig. 9).

All the feature values are normalized as per the requirement. Here, parameter values of $H_R, W_R, W_{SW}, OW_{SW}, m, n$ and d are set experimentally for the current dataset.

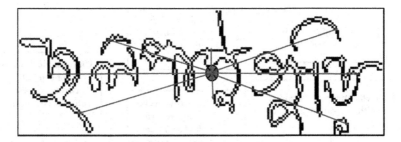

Fig. 12 Distance of farthest pixels from CG in 8-Freeman directions

2.3   Multi-objective Memetic Algorithm

MA is a meta-heuristic optimization algorithm which is an improvement on the classical GA through the inclusion of self-improvements of memes. The implementation of MA used is WFFSA (wrapper–filter feature selection) [13] which is a combination of wrapper and filter methods. The combination performs feature selection through a wrapper method while considering the influence of intrinsic properties of the features through the filter method. In particular, the filter method fine-tunes the population of GA by adding or deleting features based on ranker (here, ReliefF [15]) information. This inclusion of local search converges it to better solution much faster than GA. Therefore, in this work a multi-objective MA is preferred over GA.

The feature sets are represented as chromosomes which are encoded as binary strings where a '1' in a chromosome at position i (i.e. index of a feature in an n-element feature vector) represents that the ith feature is selected (or included) in that feature set and '0' denotes otherwise. MA begins with the creation of a random population of chromosomes. The chromosomes are evaluated using a multi-objective function described in Sect. 2.3.4. The chromosomes in the population are maintained using the elitism rule; i.e., the child chromosomes are allowed to substitute those chromosomes in the population which are inferior (according to the value of the objective function) to them. The child chromosomes are created by the operations of local search, followed by crossover and then mutation which are described in Sects. 2.3.1, 2.3.2 and 2.3.3, respectively. The flowchart of the steps is given in Fig. 13.

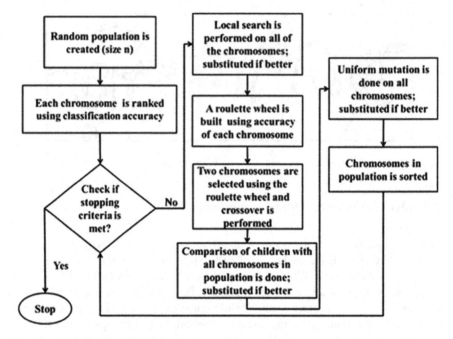

Fig. 13 Working procedure of MA

2.3.1 Local Search

Local search is the application of *Lamarkian learning* [46] which involves the creation of an ordered pair (k_1, k_2) which is used to improve each chromosome by removing k_2 least ranked features in the chromosome and add k_1 best ranked features not included in the system previously. Here, ranking is done offline once and is used multiple times by the algorithm to save time. We, in our work, have applied popularly used filter method for this called ReliefF [13]. The value of the pair, however, cannot be too large as it would lead to deletion of best-ranked features, so the values of k_1 and k_2 are upper bounded at 5% of the total number of features.

2.3.2 Crossover

This genetic operation provides the means to explore the search space. Here two-point crossover is used. The advantage of this form of crossover is the non-requirement of a probability as required for uniform crossover. Firstly, two chromosomes are selected from the population by applying the concept of a roulette wheel (discussed later in Sect. 2.3.5) using the accuracies of the chromosomes. Two random points are selected, and the bits lying between the two points are exchanged between the two parents to form two child chromosomes. The use of roulette wheel allows the parent with better classification accuracy more probability of passing on its genes (here features). The number of crossovers done in each generation is randomly selected between [2, 5]. The two-point crossover process is described in Algorithm 1.

2.3.3 Mutation

Mutation is a genetic operation which opposed to crossover provides exploitation of the search space. Uniform mutation is performed in this work. In this regard, each bit in the chromosome is flipped with probability of p. The algorithm of the uniform mutation technique is described in Algorithm 2.

Algorithm 1: Two-point crossover

Input:
p_1 and p_2 : Randomly selected parent chromosomes from population using the roulette wheel
Output:
$child_1$ and $child_2$: Children chromosome
Steps:
1. Let size of the chromosome be **n**
2. Randomly generate two natural numbers, say, n_1 and $n_2 \in [1,n]$
3. Perform two-point crossover to get $child_1$ and $child_2$ chromosomes (Swap the chromosome portion from n_1 to n_2 between p_1 and p_2)
4. return children chromosome

Algorithm 2: Uniform mutation

Input:
$P(M)$: Probability of mutation
C: A chromosome which is selected from population
Output:
Mutated chromosome
Steps:
    1. Find length of chromosome, say $L(C)$
    2. Generate a random number, say a
    3. for $(i = 1$ to $L(C))$
      {
          generate a random number $a \in [0,1]$
          if $(a<P(M))$
          {
             flip the value at position i
          }
      }
    4. return mutated chromosome

2.3.4 Objective Function

The fitness function of the multi-objective MA, used in the present work, gives more importance to the recognition accuracy (RA) than the number of features it reduces. This is because though reduction of feature dimension is necessary but that should not be achieved at the cost of the recognition ability of HWR model. The designed multi-objective function is described in Algorithm 3. Classifier used in this case is multi-layer perceptron (MLP).

Algorithm 3: Multi-objective function

```
Input:
a and b: Two chromosomes (here feature subset)
w₁: Weight for recognition accuracy (RA)
w₂: weight for feature reduction
α: RA limit which can be compromised
Output:
        better chromosome
```

$$\text{if } ((\text{mod}(RA(a) - RA(b)) > \alpha)$$

```
{
  if (RA(a) > RA(b))
     return the chromosome a
  else
     return the chromosome b
}
else
{
    define
```

$$\eta(a) = \frac{number\ of\ unused\ features\ in\ a}{Total\ number\ of\ feature\ in\ a} \quad \text{and}$$

$$\eta(b) = \frac{number\ of\ unused\ features\ in\ b}{Total\ number\ of\ feature\ in\ b}$$

$$val = \Big((w_1 \times RA(a)) + (w_2 \times \eta(a))\Big) - ((w_1 \times RA(b)) + (w_2 \times \eta(b)))$$

```
    if (val > 0)
      {
         return chromosome a as better one
      }
    else
      {
         return chromosome b as better one
      }
}
```

2.3.5 Roulette Wheel

Roulette wheel or fitness proportionate selection is a selection operator that is used to choose the parent chromosomes for performing the crossover and mutation operation. The probability (p_i) of selecting ith chromosome from the population for crossover is defined here as $p_i = \frac{RA(i)}{\sum_{i=1}^{z} RA(i)}$, where $RA(i)$ is recognition accuracy of ith chromosome which is here taken as RA and z is the size of the population. This allows the selection of fitter chromosomes to produce the children a more probable event.

2.3.6 Stopping Criteria

The number of generations generated by MA is capped at 20; i.e., the number of iterations is restricted to 20. The MA also stops or is deemed to have converged if for five continuous iterations the population undergoes no improvement; i.e., the average accuracy of the population does not change for five continuous iterations.

3 Result and Discussion

It is already mentioned that in the present work, a multi-objective MA-based feature selection technique is proposed for enhancing the performing ability of a Bangla HWR system. For experimental needs, SCF and grid-based gradient feature vectors are extracted from the word samples. A dataset containing word samples of 50 different city names is prepared here for conducting experiment. The dataset is divided into train and test sets where the test set is formed using 30 randomly selected samples per pattern class and the rest are used to form the train set. Therefore, test and train datasets contain 1500 and 6000 word samples, respectively. The train dataset is applied here for developing the word recognition model using MLP-based classifier.

3.1 Selection of Feature Dimension

To generate gradient features, each word image is hypothetically divided into 6×6 grids and then from each of the grids, a feature vector of length eight is extracted. Therefore, for gradient-based feature, a 288-dimensional feature vector is extracted from each word sample. To obtain most useful set of SCF vector in accordance with the present dataset, different experiments have been conducted by varying the parameters H_R, W_R, W_{SW}, OW_{SW}, m, n and d (see Sect. 2.2.2) and thereby different dimensional feature vectors are produced. Next, MLP-based classifier (the objective

Table 1 Various experimental outcomes observed on the current dataset by varying the parameters of SCF

Exp. #	H_R	W_R	W_{SW}	OW_{SW}	m	n	d	# of features	RA (in %)
01	32	128	32	3	2	3	4	116	77.89
02	32	128	32	3	2	3	8	148	78.46
03	64	192	32	8	2	4	8	258	85.40
04	64	192	32	4	2	3	8	258	84.53
05	64	192	32	12	2	4	8	258	86.20
06	64	192	32	16	2	4	8	258	86.20
07	64	192	32	4	2	4	8	258	85.80
08	48	160	32	8	2	4	8	216	84.80
09	**64**	**192**	**48**	**16**	**4**	**3**	**8**	**220**	**87.60**

function of MA) is used to classify the present dataset using extracted features. Some of such variations are listed in Table 1 along with corresponding recognition accuracies. The feature set, providing best result is considered here for optimization using MA. In this regard, RA, stated hereafter, is defined as

$$RA = \frac{Number\ of\ test\ samples\ correctly\ recognized}{Total\ number\ of\ test\ samples} \times 100\%$$

3.2 Recognition Results

Ample experiments have been conducted to get the better recognition result using MA-based feature selection technique on the said dataset using gradient feature, SCF and combination of these two features. The top-5 RAs, subject to parameters (see Table 2), on the test set are reported in Table 3. From this table, it is found that improved RA with reduced feature dimension has been achieved for individual as well as for combined feature vector. In the best cases, RAs are enhanced by 1.4%, 1.07% and 3.33% for gradient, SCF and combined feature vectors, respectively, in comparison with their corresponding original feature vectors. On the other hand, in these cases, reductions of feature dimensions are 15.27%, 19.09% and 39.96% for

Table 2 Parameter values used in MA

Parameter	Notation	Value
Probability of mutation	$P(M)$	0.01
Maximum size of population	z	20
Weight for RA	$w1$	1
Weight for feature reduction	$w2$	10
Maximum RA which can be compromised	α	0.5%

Table 3 Performance of the feature selection model (top-five RAs along with feature dimension are shown; best RA in each case is shown in boldface)

Feature description	Before feature selection		After feature selection			
	Size of feature vector	RA (in %)	Size of feature vector	RA (in %)	Improvement in RA (in %)	Reduction in feature dimension (in number)
Gradient	288	88.07	**258**	**89.47**	**1.40**	**30**
			267	89.40	1.33	22
			266	89.20	1.13	21
			256	89.13	1.06	32
			244	88.67	0.60	44
SCF	220	86.53	**189**	**87.60**	**1.07**	**31**
			194	87.00	0.47	26
			178	86.93	0.40	42
			190	86.73	0.20	30
			186	86.67	0.14	34
Combined	508	89.67	**324**	**93.00**	**3.33**	**184**
			325	93.00	3.33	185
			336	92.87	3.20	172
			319	92.80	3.13	189
			305	92.67	3.00	203

gradient, SCF and combined feature vectors, respectively. Impressive results justify the application of multi-objective MA for the said pattern classification problem.

3.3 Comparison with GA

Various experiments are conducted on the said dataset to compare the performing ability of MA over GA. The comparative result is shown in Table 4. In this table, top-five RAs that are found in final population by both GA and MA, and the corresponding feature dimensions are provided. The reduced feature subset rank (FSR) is estimated by its strength for recognizing test word samples. This table indicates that the MA-based feature selection approach yields better RA using less number of features in comparison with GA while considering in best RA (i.e. top choice). However, in few cases (for second and third top results), GA has generated feature subset containing less number of feature. But in these cases, it suffers from 0.6% less RA than its counterpart. Therefore, it could be concluded that MA produces better RA in all cases for getting the optimal feature set for present word recognition problem.

Table 4 Comparison between GA and MA in order to select optimal feature set for HWR problem (top-five results are tabulated). The MA and GA have been applied on combined feature vector of dimension 508 having the same configuration in genetic operation

FSR	Reduced feature dimension by		RA (in %) achieved by	
	GA	MA	GA	MA
1	**336**	**324**	**92.80**	**93.00**
2	313	325	92.40	93.00
3	315	336	92.27	92.87
4	326	319	92.27	92.80
5	318	305	92.20	92.67

4   Conclusion

Feature selection techniques assert that more information (feature) is not always helpful to design a comprehensive pattern recognition model. As a result, use of feature selection method on high-dimensional feature vector provides benefits such as identification of irrelevant features, better classification model, enhancing generalization and so on. In this paper, an objective is set to achieve the better classification accuracy with less number of features. As a case study, an MA-based feature selection technique is used for the classification of handwritten words images in a holistic approach. Two state-of-the-art feature vectors, *namely* grid-based gradient and SCF are considered here. Experimentation conducted on handwritten Bangla word samples consist of 50 different city names.

Though, the results are encouraging, still there are some scopes for the improvement of the proposed approach. Firstly, more data samples with more pattern classes are to be included in future to prove the robustness of the technique. Even, handwritten word samples from other scripts could be included to empower the present study. In addition to these, MA-based architecture could be experimented using more state-of-the-art features like topological, shape based and texture based used for HWR in holistic way. Also, local search, an important trait of MA, can also be investigated more in order to use some statistical information measuring from the excluded and included features in a chromosome.

References

1. Ghosh, M., Malakar, S., Bhowmik, S., Sarkar, R., Nasipuri, M.: Memetic algorithm based feature selection for handwritten city name recognition. In: Proceedings of First International Conference on Computational Intelligence, Communications, and Business Analytics (2017)
2. Law, M.H., Figueiredo, M.A., Jain, A.K.: Simultaneous feature selection and clustering using mixture models. IEEE Trans. Pattern Anal. Mach. Intell. **26**(9), 1154–1166 (2004)

3. Sánchez-Maroño, N., Alonso-Betanzos, A., Tombilla-Sanromán, M.: Filter methods for feature selection–a comparative study. In: International Conference on Intelligent Data Engineering and Automated Learning, pp. 178–187. Springer, Heidelberg (2007)

4. Raymer, M.L., Punch, W.F., Goodman, E.D., Kuhn, L.A., Jain, A.K.: Dimensionality reduction using genetic algorithms. IEEE Trans. Evol. Comput. **4**(2), 164–171 (2000)

5. Dey, N., Ashour, A.S., Beagum, S., Pistola, D.S., Gospodinov, M., Gospodinova, E.P., Tavares, J.M.R.: Parameter optimization for local polynomial approximation based intersection confidence interval filter using genetic algorithm: an application for brain MRI image de-noising. J. Imaging **1**(1), 60–84 (2015)

6. Karaa, W.B.A., Ashour, A.S., Sassi, D.B., Roy, P., Kausar, N., Dey, N.: MEDLINE text mining: an enhancement genetic algorithm based approach for document clustering. In: Applications of Intelligent Optimization in Biology and Medicine, pp. 267–287. Springer International Publishing (2016)

7. Xue, B., Zhang, M., Browne, W.N.: Particle swarm optimization for feature selection in classification: A multi-objective approach. IEEE Trans. Cybern. **43**(6), 1656–1671 (2013)

8. Wang, D., He, T., Li, Z., Cao, L., Dey, N., Ashour, A. S., Shi, F.: Image feature-based affective retrieval employing improved parameter and structure identification of adaptive neuro-fuzzy inference system. Neural Comput. Appl. 1–16 (2016)

9. García-Pedrajas, N., de Haro-García, A., Pérez-Rodríguez, J.: A scalable memetic algorithm for simultaneous instance and feature selection. Evol. Comput. **22**(1), 1–45 (2014)

10. Montazeri, M., Montazeri, M., Naji, H.R., Faraahi, A.: A novel memetic feature selection algorithm. In: 5th Conference on Information and Knowledge Technology (IKT), pp. 295–300. IEEE Press, New York (2013)

11. Yang, C.S., Chuang, L.Y., Chen, Y.J., Yang, C.H.: Feature selection using memetic algorithms. In: Third International Conference on Convergence and Hybrid Information Technology, vol, 1, pp. 416–423. IEEE Press, New York (2008)

12. Cateni, S., Colla, V., Vannucci, M.: A hybrid feature selection method for classification purposes. In: European Modelling Symposium. pp. 39–44. IEEE Press, New York (2014)

13. Zhu, Z., Ong, Y.S., Dash, M.: Wrapper–filter feature selection algorithm using a memetic framework. IEEE Trans. Syst. Man Cybern. Part B (Cybern.) **37**(1), 70–76 (2007)

14. Guyon, I., Elisseeff, A.: An introduction to variable and feature selection. J. Mach. Learn. Res. **3**(Mar), 1157–1182 (2003)

15. Robnik-Šikonja, M., Kononenko, I.: Theoretical and empirical analysis of ReliefF and RReliefF. Mach. Learn. **53**(1–2), 23–69 (2003)

16. Peng, H., Long, F., Ding, C.: Feature selection based on mutual information criteria of max-dependency, max-relevance and min-redundancy. IEEE Trans. Pattern Anal. Mach. Intell. **27**(8), 1226–1238 (2005)

17. Yu, L., Liu, H.: Efficient feature selection via analysis of relevance and redundancy. Journal of machine learning research. 5(Oct), 1205–1224(2004)

18. Chu, W.S., De la Torre, F., Cohn, J.F., Messinger, D.S.: A branch-and-bound framework for unsupervised common event discovery. Int. J. Comput. Vis. **123**(3), 372–391 (2017)

19. Aghdam, M.H., Ghasem-Aghaee, N., Basiri, M.E.: Text feature selection using ant colony optimization. Expert Syst. Appl. **36**(3), 6843–6853 (2009)

20. Kannan, S.S., Ramaraj, N.: A novel hybrid feature selection via symmetrical uncertainty ranking based local memetic search algorithm. Knowl. Based Syst. **23**(6), 580–585 (2010)

21. Zhu, Z., Ong, Y. S.: Memetic algorithms for feature selection on microarray data. Adv. Neural Netw. 1327–1335(2007)

22. Chen, Y., Abraham, A., Yang, B.: Feature selection and classification using flexible neural tree. Neurocomputing. **70**(1), 305–313 (2006)

23. Lee, J., Kim, D.W.: Memetic feature selection algorithm for multi-label classification. Inf. Sci. **293**, 80–96 (2015)

24. García, S., Cano, J.R., Herrera, F.: A memetic algorithm for evolutionary prototype selection: a scaling up approach. Pattern Recogn. **41**(8), 2693–2709 (2008)

25. Guillén, A., Pomares, H., González, J., Rojas, I., Valenzuela, O., Prieto, B.: Parallel multiobjective memetic RBFNNs design and feature selection for function approximation problems. Neurocomputing **72**(16), 3541–3555 (2009)
26. Hu, Z., Bao, Y., Chiong, R., Xiong, T.: Mid-term interval load forecasting using multi-output support vector regression with a memetic algorithm for feature selection. Energy. **84**, 419–431 (2015)
27. UCI Machine Learning Repository. http://archive.ics.uci.edu/ml/index.php
28. TAIR: Gene Expression—Microarray—Public Datasets. https://www.arabidopsis.org/portals/expression/microarray/microarrayDatasetsV2.jsp
29. Basu, S., Das, N., Sarkar, R., Kundu, M., Nasipuri, M., Basu, D.K.: A hierarchical approach to recognition of handwritten Bangla characters. Pattern Recogn. **42**(7), 1467–1484 (2009)
30. Barua, S., Malakar, S., Bhowmik, S., Sarkar, R., Nasipuri, M.: Bangla handwritten city name recognition using gradient based feature. In: 5th International Conference on Frontiers of Intelligent Computing: Theory and Applications (FICTA), pp. 343–352. Springer, Singapore (2017)
31. Malakar, S., Sharma, P., Singh, P.K., Das, M., Sarkar, R., Nasipuri, M.: A holistic approach for handwritten hindi word recognition. Int. J. Comput. Vis. Image Process. (IJCVIP). **7**(1), 59–78 (2017)
32. Bhowmik, S., Polley, S., Roushan, M.G., Malakar, S., Sarkar, R., Nasipuri, M.: A holistic word recognition technique for handwritten Bangla words. Int. J. Appl. Pattern Recognit. **2**(2), 142–159 (2015)
33. Bhowmik, S., Malakar, S., Sarkar, R., Nasipuri, M.: Handwritten bangla word recognition using elliptical features. In: International Conference on Computational Intelligence and Communication Networks (CICN), pp. 257–261. IEEE Press, New York (2014)
34. Bhowmik, S., Roushan, M. G., Sarkar, R., Nasipuri, M., Polley, S., Malakar, S.: Handwritten Bangla word recognition using hog descriptor. In: Fourth International Conference of Emerging Applications of Information Technology (EAIT), pp. 193–197, IEEE Press, New York (2014)
35. Pal, U., Roy, K., Kimura, F.: A lexicon-driven handwritten city-name recognition scheme for Indian postal automation. IEICE Trans. Inf. Syst. **92**(5), 1146–1158 (2009)
36. Roy, K., Vajda, S., Pal, U., Chaudhuri, B. B.: A system towards Indian postal automation. In: Ninth International Workshop on Frontiers in Handwriting Recognition, pp. 580–585. IEEE Press, New York (2004)
37. Dzuba, G., Filatov, A., Gershuny, D., Kil, I., Nikitin, V.: Check amount recognition based on the cross validation of courtesy and legal amount fields. Int. J. Pattern Recognit Artif Intell, **11**(04), 639–655 (1997)
38. Roy, P.P., Bhunia, A.K., Das, A., Dhar, P., Pal, U.: Keyword spotting in doctor's handwriting on medical prescriptions. Expert Syst. Appl. **76**, 113–128 (2017)
39. Languages with at least 50 million first-language speakers. Retrieved from Summary by language size Ethnologue. https://www.ethnologue.com/statistics/size
40. Tamen, Z., Drias, H., Boughaci, D.: An efficient multiple classifier system for Arabic handwritten words recognition. Pattern Recognit. Lett. **93**, 123–132 (2017)
41. Hemalatha, S., Anouncia, S.M.: Unsupervised Segmentation of Remote Sensing Images using FD Based Texture Analysis Model and ISODATA. Int. J. Ambient Comput. Intell. (IJACI) **8**(3), 58–75 (2017)
42. Dalal, N., Triggs, B.: Histograms of oriented gradients for human detection. In: IEEE Computer Society Conference on Computer Vision and Pattern Recognition, vol. 1, pp. 886–893. IEEE (2005)
43. Dey, N., Ashour, A.S., Hassanien, A.E.: Feature detectors and descriptors generations with numerous images and video applications: a recap. In: Feature Detectors and Motion Detection in Video Processing, IGI Global, pp. 36–65 (2017)
44. Gonzalez, R.C.: Digital Image Processing. Pearson Education, India (2009)

45. Freeman, H.: On the encoding of arbitrary geometric configurations. IRE Trans. Electron. Comput. **10**, 260–268 (1961)
46. Ong, Y.S., Keane, A.J.: Meta-Lamarckian learning in memetic algorithms. IEEE Trans. Evol. Comput. **8**(2), 99–110 (2004)

A Column-Wise Distance-Based Approach for Clustering of Gene Expression Data with Detection of Functionally Inactive Genes and Noise

Girish Chandra and Sudhakar Tripathi

Abstract Due to uncertainty and inherent noise present in gene expression data, clustering of the data is a challenging task. The common assumption of many clustering algorithms is that each gene belongs to a cluster. However, few genes are functionally inactive, i.e. not participate in any biological process during experimental conditions and should be segregated from clusters. Based on this observation, a clustering method is proposed in this article that clusters co-expressed genes and segregates functionally inactive genes and noise. The proposed method formed a cluster if the difference in expression levels of genes with a specified gene is less than a threshold t in each experimental condition; otherwise, the specified gene is marked as functionally inactive or noise. The proposed method is applied on 10 yeast gene expression data, and the result shows that it performs well over existing one.

Keywords Gene expression data · Clustering · Data mining

1 Introduction

DNA microarray technology provides a great opportunity to simultaneously monitor the expression levels of thousands of genes involved in many biological processes. The expression levels for a set of genes are measured under different

This paper is a revised and expanded version of a paper entitled "A new approach for clustering gene expression data" *presented at* Computational Intelligence, Communications, and Business Analytics (CICBA-2017) organised by Calcutta Business School on March 24–25, 2017.

G. Chandra (✉) · S. Tripathi
Department of Computer Science & Engineering, National Institute of Technology Patna, Patna, Bihar, India
e-mail: gcchandra440@gmail.com

S. Tripathi
e-mail: p.stripathi@gmail.com

© Springer Nature Singapore Pte Ltd. 2019
J. K. Mandal et al. (eds.), *Advances in Intelligent Computing*,
Studies in Computational Intelligence 687,
https://doi.org/10.1007/978-981-10-8974-9_7

experimental conditions or samples. Analysis of the expression levels of the genes, measured by DNA microarray chip, is helpful in the study of the functions of genes.

A gene expression data contain expression levels of a set of genes measured during different experimental conditions. A gene expression data are represented in the form of a matrix where a row represents a gene vector and a column represents an experimental condition.

It has been observed that genes having similar expression levels, i.e. co-expressed, of a gene expression data are also involved in similar biological functions. To identify these co-expressed genes, clustering plays an important role. Clustering is a process of dividing genes of a gene expression data into several groups, called as clusters, such that genes in a cluster are similar in nature and genes of different clusters are dissimilar in nature.

There are many clustering algorithms that have been successfully implemented on gene expression data. The algorithms such as k-means [1], fuzzy c-means [2], hierarchal [3], SOM [4], CLICK [5] and SiMM-TS [6] work well in identifying similar expression gene clusters. These algorithms assign each gene to a cluster. However, it has been observed that a few genes of gene expression data set are functionally inactive during some experimental conditions and do not participate in any biological activities [7, 8]. That is, these functionally inactive genes should not be a part of any clusters. Therefore, there is a necessity of an algorithm that divides genes of gene expression data set into two groups: one is co-expressed gene clusters, and another is functionally inactive gene set.

The aforementioned problem is identified in this article, and a method is proposed that clusters co-expressed genes and also identifies functionally inactive genes. The method clusters the genes, if they are within distance threshold t with respect to a specified gene in each experimental condition. If none of the genes is within distance threshold t with respect to the specified gene in each experimental condition, then the specified gene is classified as a functionally inactive gene or noise. To validate the proposed method, an experimental analysis has been made on 10 yeast gene expression data. To validate the experimental result, clustering validation indices such as Dunn [9, 10], Silhouette [10, 11], Connectivity [12] have been used. And also a comparative analysis has been made with k-means, hierarchal and SOM clustering procedure. The comparative result shows the capabilities of the proposed method over the existing algorithms.

The paper is organized into four sections. After Introduction, Sect. 2 describes the proposed method. Section 3 contains results of the proposed method on 10 yeast gene expression data, description of cluster validation indices, and comparison of the proposed method with existing algorithms. Section 4 contains conclusion of this work.

2 Proposed Method

In this section, our main objective is to enhance the clustering process for gene expression data. In recent study, it seems that no one algorithm identified the noise or non-functional genes in clustering. But, what is the importance of noisy or non-functional genes in clustering? To overcome this limitation of clustering, we proposed a new clustering method, which clusters co-expressed genes and segregates functionally inactive gene and noise. The proposed method clusters the genes based on column-wise distance with specific genes. The genes having distance less than a threshold t with respect to a specific gene in each experimental conditions are assigned to a cluster. If no genes found that are less than parameter t in each experimental conditions, then the specific gene is marked as functionally inactive or noise. The detailed procedure of the proposed method is given below:

For a given gene expression data set $GE = \{w_{ij} | 1 \leq i \leq n; 1 \leq j \leq m\}$ of n genes and m experimental conditions, where w_{ij} is the expression level of the gene $g_i \in G = \{g_1, g_2, ..., g_n\}$ under experimental condition $e_i \in E = \{e_1, e_2, ..., e_m\}$, two sets, named as unclustered gene set U and marked gene set M, are used. The uncluster gene set U holds the unclustered genes. Initially, U has all the genes of the gene expression data set. The marked gene set U contains the unclustered genes (except marked gene g_i) that have the possibility to form a new cluster.

The method starts with selecting a gene g_i randomly from the unclusted gene set U. After that, all unclustered genes, except g_i, are assigned to the marked gene set M. We define a single column (experimental condition e_j) distance d between a pair of genes g_i and g_k as

$$d = |w_{ij} - w_{kj}| \tag{1}$$

This distance d is computed for all genes of marked gene set M with gene g_i. We define a parameter threshold t that is used to remove genes from the marked gene set M. It is observed that a part of genes of the marked gene set M have distance d more than threshold t and these genes are removed from the marked gene set M. The size of the removed genes is depending both on the distribution of expression levels in the gene expression data set and the parameter threshold t. These two factors also decide the size of the clusters. High-range expression levels and low threshold t yield smaller size clusters. On the other hand, high threshold t yields larger size clusters.

The distance d calculation and removal of genes from marked gene set M is done for each experimental condition one by one. Sometimes, marked gene set M becomes empty either in the middle of experimental conditions or after completing all experimental conditions. In this case, the gene g_i has no genes in the column-wise t-neighbourhood and the gene g_i is assumed to be either functionally inactive or noise. In the second case, when the marked gene set M is non-empty for all experimental conditions, then the genes of marked gene set M with gene g_i formed a new cluster. After formation of a new cluster, the genes of the cluster, or gene g_i (in the case of noise), are removed from the unclustered gene set U. After updating unclustered gene set U, aforementioned steps are repeated till unclustered gene set not empty. With the formation of new clusters, the unclustered genes set U reduces and next cluster formation requires less time as compared to previously formed cluster(s).

The algorithm of the proposed method is as follows:

Algorithm 1: Gene Clustering algorithm based on column wise distance

---

1. *Input: GE: Gene expression data*
2. *Output: c: Clusters*
3. *   Nf: Noise/functionally inactive genes*
4. *c ←1*
5. *U ← unclustered genes*
6. *M ← marked genes*
7. **While** $U \neq \phi$ **do**
8.   **for each** *unclustered gene $g_i \in U$* **do**
9.   *M←U-{g_i}*
10.   **for each** *experiment condition e_j & marked gene*
11. $g_k \in M$ **do**
12.   **if** $|w_{ij} - w_{kj}| > t$ **then**
13.   *M←M - g_k*
14.   **end if**
15.   **end for**
16.   **if** $M = \phi$ **then**
17.   *label(g_i)←Nf*
18.   **else**
19.   *label(M ∪ g_i) ← c*
20.   *c←c+1*
21.   **end if**
22.   *update U*
23.   **end for**
24. **end do**

---

3   Implementation and Analysis

In this section, our main objective is to validate the proposed clustering method. To do so, we applied the proposed method on 10 yeast gene expression data set. These data are previously used in Maji and Paul [13] and publicly available at NCBI Web site (https://www.ncbi.nlm.nih.gov/sites/GDSbrowser). The expression levels of

Table 1 Names of ten yeast gene expression data set	GDS2713	GDS2712	GDS759	GDS1013	GDS1550
	GDS1611	GDS2002	GDS2003	GDS2196	GDS608

these gene expression data sets are normalized such that the mean and variance is 0 and 1, respectively. The description of the data set is listed in Table 1.

3.1 Cluster Validation Index

To evaluate and validate the clusters generated by the proposed method, the Dunn index [9, 10], Silhouette [10, 11] and Connectivity [12] have been used in this study and also used to compare the proposed method with the existing one.

Dunn Index.
It is the ratio of the smallest measured distance between two objects of any two clusters with the largest measured distance of two objects of a cluster. The formula to calculate Dunn index is

$$Dunn = \min_{1 \leq i \leq n, 1 \leq j \leq n, i \neq j} \left(\frac{d(i,j)}{max_{1 \leq k \leq n} d'(k)} \right) \tag{2}$$

where the number of obtained clusters is n, the distance between ith cluster centre and jth cluster centre is $d(i,j)$, and the maximum distance measured between two objects of the kth cluster is $d'(k)$. The value of Dunn index lies between 0 to ∞. The value of Dunn index should be maximized.

Connectivity.
Connectivity is defined as

$$conn = \sum_{i=1}^{n} \sum_{j=1}^{k} W_{ij} \tag{3}$$

where the number of objects in the data set is n. The value of W_{ij} is either 0 or $1/j$. If object i and jth nearest neighbour of object i are the member of one cluster, then W_{ij} is 0, otherwise $1/j$, and k is a parameter to consider nearest neighbour for an object. The Connectivity index value lies between 0 to ∞. This index value should be minimized.

Silhouette.
Silhouette is calculated by taking an average of Silhouette value of each object. Silhouette of an object i is defined as

$$S(i) = \frac{b_i - a_i}{max(b_i, a_i)} \qquad (4)$$

where a_i is the average distance of object i with all other objects of the same cluster and b_i is the average distance of object i with objects of the nearest neighbour clusters.

The range of Silhouette is from −1 to 1. Smaller or negative Silhouette value of an object indicates that the assignment of the object to a cluster is worst, and higher or positive Silhouette value of an object indicates that the object is well clustered.

3.2 Implementation Platform

The implementation of the proposed method is on R. To evaluate and validate the results of clustering algorithms, two R packages, clValid [12] and clv [14], are used.

3.3 Results

The proposed method is applied on each yeast gene expression data set with different values of parameter threshold t, staring from 0.1 and up to that value where clusters are obtained, with an interval of 0.1. It is commonly observed that for a smaller threshold t, a large number of clusters are obtained and size of a cluster is small. A lot of functionally inactive genes or noises are also obtained. In the case of large threshold t, the behaviour of the proposed method is opposite. Small number of clusters is obtained, more number of genes assigns to the clusters, and a very small number of genes are noise. The Dunn index, Connectivity, and Silhouette are calculated at different threshold t for each yeast gene expression data, and best 10 results indicated by these indices are listed in descending order in Table 2, Table 3 and Table 4, respectively.

The clustering algorithms, k-means, hierarchal and SOM, are also applied on these yeast gene expression data set with a different number of clusters (parameter). The optimal number of clusters indicated by Dunn index, Connectivity and Silhouette with optimal values of these indices is shown in Table 2, Table 3 and Table 4, respectively.

The observations from the Table 2 are as follows:

GDS608: The best clusters, by the proposed method, are obtained at t = 0.3 where the Dunn index value is maximum i.e. 0.285644. However, number of genes identified in clusters are 492 which is very small, and it is not desirable. Therefore, we consider some more clusters having a significant number of genes in clusters with moderate value of Dunn indices. The Dunn indices at t 0.7, 5.5, 0.6, 0.8 have 2813, 3997, 2385, 3142 genes, respectively, in clusters, and we can consider these clusters.

Table 2 Best ten Dunn index values obtained by the proposed method and optimal Dunn index obtained by k-means, hierarchal and SOM applied on each ten yeast gene expression data

Data set	Proposed method				k-means		Hierarchal		SOM	
	Threshold t	Dunn	Cluster	No of genes obtained in clusters	Cluster	Dunn	Cluster	Dunn	Cluster	Dunn
GDS608	0.3	0.285644	199	492	15	0.093422	3	0.1460932	5	0.102602
	0.4	0.187941	382	1208						
	0.7	0.105381	454	2813						
	5.5	0.103102	2	3997						
	0.6	0.097052	453	2385						
	0.8	0.088066	428	3142						
	4.8	0.077825	5	3997						
	4.2	0.074113	9	3997						
	4.7	0.068069	6	3997						
	0.9	0.067396	368	3369						
GDS759	0.5	0.359006	114	267	19	0.1207715	2	0.2140476	7	0.175429
	0.6	0.27915	241	724						
	0.7	0.223064	303	1169						
	6.9	0.204023	2	2499						
	0.8	0.179728	301	1513						
	0.9	0.146184	276	1789						
	6.4	0.146084	2	2499						
	6.7	0.146084	2	2500						
	1	0.130816	235	1963						
	6.5	0.125339	3	2499						

(continued)

Table 2 (continued)

Data set	Proposed method				k-means		Hierarchal		SOM	
	Threshold t	Dunn	Cluster	No of genes obtained in clusters	Cluster	Dunn	Cluster	Dunn	Cluster	Dunn
GDS1013	1.2	0.117976	30	4537	2	0.1059739	2	0.0992662	2	0.111751
	3.9	0.103326	12	4632						
	5.7	0.101221	7	4634						
	4.6	0.097653	9	4630						
	6.5	0.095287	7	4636						
	1.5	0.093821	33	4581						
	6.2	0.090958	7	4637						
	6	0.089579	6	4636						
	0.9	0.088284	35	4494						
	3.1	0.087699	11	4620						
GDS1550	5.4	0.086581	3	4635	4	0.06830188	2	0.0751763	5	0.072108
	4.5	0.079623	4	4635						
	4.1	0.077853	5	4636						
	6.1	0.071516	3	4636						
	6	0.07095	3	4636						
	6.4	0.070334	3	4636						
	6.6	0.064581	3	4636						
	5.1	0.060093	3	4636						
	5.6	0.058112	3	4635						
	1.7	0.051927	13	4628						

(continued)

Table 2 (continued)

Data set	Proposed method				k-means		Hierarchal		SOM	
	Threshold t	Dunn	Cluster	No of genes obtained in clusters	Cluster	Dunn	Cluster	Dunn	Cluster	Dunn
GDS1611	0.3	0.185825	413	1123	6	0.15644	5	0.13429	9	0.147823
	0.4	0.177692	594	2673						
	0.6	0.085569	330	4095						
	0.7	0.080782	233	4330						
	0.5	0.066397	493	3629						
	1	0.055009	83	4562						
	0.9	0.046346	120	4523						
	0.8	0.044487	166	4460						
	1.2	0.040501	50	4602						
	1.4	0.038835	30	4619						
GDS2002	0.4	0.469003	78	184	7	0.187941	4	0.22632	4	0.162079
	0.5	0.316724	250	620						
	0.6	0.241617	503	1599						
	0.7	0.171659	610	2636						
	0.8	0.152503	553	3346						
	11	0.148596	2	5468						
	1	0.140951	388	4200						
	0.9	0.140508	480	3857						
	9.4	0.128488	3	5467						
	7	0.126924	8	5464						

(continued)

Table 2 (continued)

Data set	Proposed method				k-means		Hierarchal		SOM	
	Threshold t	Dunn	Cluster	No of genes obtained in clusters	Cluster	Dunn	Cluster	Dunn	Cluster	Dunn
GDS2003	0.3	0.775958	32	78	9	0.105381	3	0.12387	3	0.11779
	0.4	0.611162	45	112						
	0.5	0.372007	138	338						
	0.6	0.233947	334	1002						
	0.7	0.192353	465	1786						
	0.9	0.170306	414	2796						
	0.8	0.152303	443	2387						
	1	0.145779	337	3087						
	5.6	0.129735	11	4311						
	9.6	0.127694	3	4310						
GDS2196	8.2	0.106462	3	4636	3	0.103102	3	0.140238	5	0.106089
	8.6	0.086605	3	4636						
	8.8	0.081508	3	4637						
	8.9	0.081508	3	4637						
	4.1	0.078717	5	4635						
	8.4	0.076421	3	4637						
	8.5	0.076421	3	4637						
	8.7	0.072941	4	4637						
	7.6	0.068613	4	4636						
	9	0.068613	3	4637						

(continued)

Table 2 (continued)

Data set	Proposed method				k-means		Hierarchal		SOM	
	Threshold t	Dunn	Cluster	No of genes obtained in clusters	Cluster	Dunn	Cluster	Dunn	Cluster	Dunn
GDS2712	6.8	0.094577	3	7825	5	0.07531263	3	0.090826	3	0.072103
	6.6	0.090911	3	7824						
	6.9	0.089973	3	7825						
	7	0.08972	3	7825						
	6.7	0.087656	2	7824						
	0.1	0.068836	280	4672						
	3.5	0.063924	15	7816						
	3.8	0.058231	12	7818						
	5.7	0.049223	7	7825						
	6	0.048484	5	7824						
GDS2713	0.3	0.046361	254	6880	5	0.04409552	5	0.0520783	5	0.041273
	0.1	0.046271	313	4513						
	4.4	0.046219	4	7822						
	4.9	0.041468	3	7824						
	4.1	0.041166	3	7824						
	2.4	0.040021	14	7819						
	2	0.039938	25	7811						
	3.4	0.037466	7	7823						
	3.6	0.037158	5	7822						
	2.9	0.035719	11	7820						

Table 3 Best ten Connectivity values obtained by the proposed method and optimal Connectivity obtained by k-means, hierarchal and SOM applied on each ten yeast gene expression data

Data set	Proposed method				k-means		Hierarchal		SOM	
	Threshold t	Connectivity	Cluster	No of genes obtained in clusters	Cluster	Optimal Connectivity	Cluster	Optimal Connectivity	Cluster	Optimal Connectivity
GDS608	5.5	6.569048	2	3997	8	51.025081	2	11.045342	3	15.71671
	4.9	17.51111	4	3997						
	5.6	34.0131	3	3997						
	4.5	50.00079	5	3996						
	5.1	52.2123	5	3998						
	5.4	54.61508	5	3999						
	4.8	59.2123	5	3997						
	4.2	82.90992	9	3997						
	5	85.85595	2	3998						
	4.7	107.9921	6	3997						
GDS759	6.9	5.357937	2	2499	2	16.523091	2	5.1440476	2	8.09667
	6.4	8.286905	2	2499						
	6.7	10.63254	2	2500						
	6.5	15.35198	3	2499						
	6.8	18.05198	2	2499						
	6.6	21.15833	3	2499						
	6.2	22.84405	3	2498						
	6	27.99087	3	2499						
	5.7	29.9369	4	2499						
	6.3	30.87778	3	2499						

(continued)

Table 3 (continued)

Data set	Proposed method				k-means		Hierarchal		SOM	
	Threshold t	Connectivity	Cluster	No of genes obtained in clusters	Cluster	Optimal Connectivity	Cluster	Optimal Connectivity	Cluster	Optimal Connectivity
GDS1013	6.7	51.20635	5	4635	2	5.5420635	2	3.8579365	2	71.2333
	7	52.31111	5	4636						
	6.1	54.31389	5	4635						
	6.9	56.20635	4	4636						
	6.5	58.44563	7	4636						
	6.4	60.13889	5	4635						
	6.8	60.82976	5	4636						
	6.6	61.29325	6	4635						
	5.4	63.43492	7	4635						
	6.3	64.22778	5	4635						
GDS1550	6.4	15.97222	3	4636	2	6.0507365	2	3.857365	4	10.28331
	5.4	18.40675	3	4635						
	7	18.82738	3	4637						
	4.5	21.08254	4	4635						
	4.7	21.18849	3	4635						
	6.8	21.60357	3	4637						
	5.1	22.17897	3	4636						
	6.6	22.69365	3	4636						
	6.2	23.54524	3	4636						
	6.9	23.59484	3	4636						

(continued)

Table 3 (continued)

Data set	Proposed method				k-means		Hierarchal		SOM	
	Threshold t	Connectivity	Cluster	No of genes obtained in clusters	Cluster	Optimal Connectivity	Cluster	Optimal Connectivity	Cluster	Optimal Connectivity
GDS1611	3.2	9.370238	2	4637	5	46.340912	3	34.75688	6	56.71673
	2.9	54.49286	2	4636						
	3	186.7663	3	4636						
	2.5	193.8452	3	4636						
	2.7	195.148	2	4636						
	3.1	296.8353	2	4637						
	2.8	392.3825	4	4636						
	2.6	413.9837	3	4637						
	2.2	448.5405	5	4633						
	2.4	450.2317	5	4635						
GDS2002	11	5.426191	2	5468	6	18.08234	3	17.06654	5	21.02233
	10.2	11.09881	2	5467						
	10.5	11.62738	2	5469						
	10.8	11.62738	2	5469						
	10.9	11.62738	2	5469						
	10	13.37143	2	5467						
	10.3	14.0131	2	5469						
	10.7	14.27024	2	5468						
	10.1	15.98373	2	5467						
	10.6	15.98373	2	5467						

(continued)

Table 3 (continued)

Data set	Proposed method				k-means		Hierarchal		SOM	
	Threshold t	Connectivity	Cluster	No of genes obtained in clusters	Cluster	Optimal Connectivity	Cluster	Optimal Connectivity	Cluster	Optimal Connectivity
GDS2003	9.5	8.527381	2	4309	3	12.56745	5	16.050432	4	28.340561
	9.4	8.527381	2	4310						
	9.8	8.527381	2	4311						
	9.7	14.08492	3	4311						
	9.6	16.08532	3	4310						
	8.9	20.32937	3	4309						
	9.3	20.32937	3	4312						
	9.1	21.72698	3	4310						
	8.8	24.88611	3	4309						
	8.6	28.15992	3	4308						
GDS2196	8.8	10.70873	3	4637	3	11.04533	3	13.045534	4	36.38937
	8.9	10.70873	3	4637						
	9	11.56825	3	4637						
	7.9	13.12103	3	4636						
	8.5	13.98492	3	4637						
	8.6	14.44841	3	4636						
	7.7	14.70119	3	4637						
	7.8	14.86706	3	4637						
	7.1	14.87659	3	4637						
	6.9	14.90675	3	4636						

(continued)

Table 3 (continued)

Data set	Proposed method				k-means		Hierarchal		SOM	
	Threshold t	Connectivity	Cluster	No of genes obtained in clusters	Cluster	Optimal Connectivity	Cluster	Optimal Connectivity	Cluster	Optimal Connectivity
GDS2712	6.7	70.65595	2	7824	2	56.1805556	2	14.653492	2	40.98331
	6.4	78.0869	3	7824						
	6.3	81.85714	4	7824						
	6.6	94.22262	3	7824						
	5.2	97.48135	5	7821						
	6.8	97.77937	3	7825						
	6.5	99.18254	4	7825						
	6.9	100.623	3	7825						
	7	104.4175	3	7825						
	6.1	113.2163	4	7823						
GDS2713	5	143.719	3	7824	4	59.419440	3	22.55518	3	41.15475
	4.1	145.9012	3	7824						
	4.9	146.6238	3	7824						
	4.4	156.2671	4	7822						
	3.4	239.0115	7	7823						
	4.7	248.3119	4	7823						
	4.3	276.5532	3	7823						
	4.8	293.152	4	7824						
	4.5	294.1845	3	7823						
	3.6	301.4266	5	7822						

Table 4 Best ten Silhouette values obtained by the proposed method and optimal Silhouette index obtained by k-means, hierarchal and SOM applied on each ten yeast gene expression data

Data set	Propcsed method				k-means		Hierarchal		SOM	
	Threshold t	Silhouette	Cluster	No of genes obtained in clusters	Cluster	Silhouette	Cluster	Silhouette	Cluster	Silhouette
GDS608	5	0.443244	2	3998	10	0.360764	5	0.33456	5	0.33456
	5.5	0.415159	2	3997						
	5.6	0.364708	3	3997						
	4.8	0.350789	5	3997						
	4.9	0.350315	4	3997						
	4.6	0.331767	4	3997						
	4.5	0.329987	5	3996						
	5.4	0.327507	5	3999						
	5.3	0.323643	3	3998						
	5.2	0.323636	6	3999						
GDS759	6.9	0.701075	2	2499	2	0.4239139	2	0.653950	2	0.617404
	6.4	0.665596	2	2499						
	6.7	0.657885	2	2500						
	6.5	0.655838	3	2499						
	6.8	0.648067	2	2499						
	6.3	0.609826	3	2499						
	6.6	0.599442	3	2499						
	6	0.580986	3	2499						
	6.1	0.57835	2	2498						
	5.5	0.534962	4	2499						

(continued)

Table 4 (continued)

Data set	Proposed method				k-means		Hierarchal		SOM	
	Threshold t	Silhouette	Cluster	No of genes obtained in clusters	Cluster	Silhouette	Cluster	Silhouette	Cluster	Silhouette
GDS1013	6.7	0.928403	5	4635	4	0.9260471	5	0.9122757	5	0.870345
	6.9	0.925931	4	4636						
	6.8	0.925676	5	4636						
	6.1	0.925428	5	4635						
	6.4	0.924621	5	4635						
	6.3	0.92371	5	4635						
	5.6	0.918428	5	4633						
	5.7	0.912976	7	4634						
	6.2	0.907884	7	4637						
	7	0.90532	5	4636						
GDS1550	6.2	0.923376	3	4636	3	0.90275666	3	0.9162063	3	0.902453
	7	0.921821	3	4637						
	6.8	0.920066	3	4637						
	6.4	0.919995	3	4636						
	6.6	0.919822	3	4636						
	6.3	0.919821	3	4636						
	5.3	0.918875	3	4636						
	5.7	0.91879	3	4636						
	6.7	0.918348	3	4636						
	6.1	0.91796	3	4636						

(continued)

Table 4 (continued)

Data set	Proposed method				k-means		Hierarchal		SOM	
	Threshold t	Silhouette	Cluster	No of genes obtained in clusters	Cluster	Silhouette	Cluster	Silhouette	Cluster	Silhouette
GDS1611	3.2	0.573747	2	4637	3	0.553420	2	0.68454	3	0.560201
	2.9	0.49911	2	4636						
	2.7	0.468892	2	4636						
	3.1	0.463809	2	4637						
	2.5	0.270175	3	4636						
	2.8	0.267849	4	4636						
	3	0.215157	3	4636						
	2.6	0.19024	3	4637						
	2.3	0.152544	6	4635						
	2.4	0.124947	5	4635						
GDS2002	11	0.889162	2	5468	3	0.767451	4	0.841392	4	0.734561
	10.5	0.866513	2	5469						
	10.8	0.866513	2	5469						
	10.9	0.866513	2	5469						
	10.7	0.862158	2	5468						
	10.3	0.85845	2	5469						
	10.6	0.856437	2	5467						
	10.1	0.856373	2	5467						
	10.4	0.855977	2	5469						
	10	0.846804	2	5467						

(continued)

Table 4 (continued)

Data set	Proposed method				k-means		Hierarchal		SOM	
	Threshold t	Silhouette	Cluster	No of genes obtained in clusters	Cluster	Silhouette	Cluster	Silhouette	Cluster	Silhouette
GDS2003	0.3	0.945976	32	78	3	0.805243	2	0.812302	2	0.775654
	9.5	0.814348	2	4309						
	9.4	0.814319	2	4310						
	9.8	0.814022	2	4311						
	9.6	0.731596	3	4310						
	8.9	0.725754	3	4309						
	9.3	0.724967	3	4312						
	9.7	0.721901	3	4311						
	0.4	0.719234	45	112						
	9.1	0.706973	3	4310						
GDS2196	8.2	0.952399	3	4636	5	0.949033	3	0.965649	5	0.868301
	9	0.95041	3	4637						
	8.3	0.950284	3	4636						
	8.8	0.949927	3	4637						
	8.9	0.949927	3	4637						
	8.5	0.948992	3	4637						
	8.4	0.948535	3	4637						
	7.9	0.948009	3	4636						
	8	0.947792	3	4637						
	8.1	0.947792	3	4637						

(continued)

Table 4 (continued)

Data set	Proposed method				k-means		Hierarchal		SOM	
	Threshold t	Silhouette	Cluster	No of genes obtained in clusters	Cluster	Silhouette	Cluster	Silhouette	Cluster	Silhouette
GDS2712	6.7	0.845303	2	7824	5	0.67241096	5	0.7920550	5	0.75734
	6.9	0.840101	3	7825						
	6.4	0.839859	3	7824						
	6.8	0.836918	3	7825						
	6.6	0.834476	3	7824						
	7	0.833573	3	7825						
	6	0.773461	5	7824						
	5.8	0.766377	5	7824						
	3.8	0.740121	12	7818						
	6.2	0.707578	4	7825						
GDS2713	4.9	0.754239	3	7824	12	0.637414	5	0.7298310	5	0.716029
	5	0.736408	3	7824						
	4.4	0.73326	4	7822						
	4.5	0.730555	3	7823						
	4.2	0.726078	5	7824						
	4.6	0.722985	4	7823						
	3.7	0.71174	8	7824						
	4.1	0.708088	3	7824						
	3.9	0.705807	6	7823						
	3.1	0.694722	7	7823						

The Dunn indices of the proposed method at t 0.3, 0.4 are greater than optimal Dunn index of the hierarchal algorithm. It is also greater at t 0.3, 0.4, 0.7, 5.5 in comparison with optimal Dunn index values of k-means and SOM. This shows that the proposed method performs better in comparison with k-means, hierarchal and SOM for this data set.

GDS759: The maximum Dunn index value for the proposed method is 0.359006 obtained at t = 0.5. Similar to the results of GDS608, it only contains 267 in clusters. Therefore, we can consider some more clusters with a significant number of genes in cluster shown in the rows of GDS759 of Table 2.

The Dunn indices of the proposed method at t 0.5, 0.6 and 0.7 are greater than optimal Dunn index of the hierarchal algorithm. It is also greater at t 0.5, 0.6, 0.7, 6.9 and 0.8 in comparison with optimal Dunn index values of k-means and SOM.

GDS1013: The maximum Dunn index value for the proposed method is 0.117976 obtained at t = 1.2, and 4537 genes are identified in 30 clusters.

The Dunn index of the proposed method at t 1.2 is greater than optimal Dunn index value of k-means and SOM algorithm. It is also greater at t 1.2, 3.9 and 5.7 in comparison with optimal Dunn index values of the hierarchal algorithm.

GDS1550: The best clusters are obtained at t = 5.4 where the Dunn index value is maximum i.e. 0.086581. The Dunn index values of the proposed method at t 5.4, 4.5 and 4.1 are greater than optimal Dunn index value of k-means, hierarchal and SOM.

GDS1611: The best clusters are obtained at t = 0.3 where the Dunn index value is maximum i.e. 0.185825. The Dunn index values of the proposed method at t 0.3 and 0.4 are greater than optimal Dunn index value of k-means, hierarchal and SOM.

GDS2002: The best clusters are obtained at t = 0.4 where the Dunn index value is maximum i.e. 0.469003. The Dunn index values of the proposed method at t 0.4, 0.5 and 0.6 are greater than optimal Dunn index value of k-means, hierarchal and SOM.

GDS2003: The best clusters are obtained at t = 0.3 where the Dunn index value is maximum i.e. 0.775958. All 10 Dunn index values of the proposed method are greater than optimal Dunn index value of k-means, hierarchal and SOM.

GDS2196: The best clusters are obtained at t = 8.2 where the Dunn index value is maximum i.e. 0.106462. The Dunn index value of the proposed method at t 8.2 is greater than optimal Dunn index value of k-means and SOM.

GDS2712: The best clusters are obtained at t = 6.8 where the Dunn index value is maximum i.e. 0.094577. The Dunn index values of the proposed method at t 6.8 and 6.6 are greater than optimal Dunn index value of k-means, hierarchal and SOM.

GDS2713: The best clusters are obtained at t = 0.3 where the Dunn index value is maximum i.e. 0.046361. The Dunn index values of the proposed method at t 0.3, 0.1 and 4.4 are greater than optimal Dunn index value of k-means and SOM.

The above observations, based on Dunn index, prove that the proposed method performs better on the data set GDS608, GDS759, GDS1013, GDS1550, GDS1611, GDS2002, GDS2003, GDS2712 and GDS2713 in comparison with k-means, hierarchal and SOM. For the data set GDS2196, the proposed method performs better than k-means and SOM.

The observations from the Table 3 are as follows:

GDS608: The best clusters are obtained at t = 5.5 where the Connectivity is minimum; i.e., 6.569048 and 3997 genes are obtained in two clusters. The Connectivity of the proposed method at t 5.5 is smaller than optimal Connectivity of k-means and SOM.

GDS759: The best clusters are obtained at t = 6.9 where the Connectivity is minimum; i.e., 5.357937 and 2499 genes are obtained in two clusters. The Connectivity of the proposed method at t 6.9 is smaller than optimal Connectivity of k-means and SOM.

GDS1013: The best clusters are obtained at t = 6.7 where the Connectivity is minimum; i.e., 51.20635 and 4635 genes are obtained in five clusters. All Connectivity values of the proposed method are smaller than optimal Connectivity of SOM.

GDS1550: The best clusters are obtained at t = 6.4 where the Connectivity is minimum; i.e., 15.97222 and 4636 genes are obtained in three clusters.

GDS1611: The best clusters are obtained at t = 3.2 where the Connectivity is minimum; i.e., 9.370238 and 4637 genes are obtained in two clusters. The Connectivity of the proposed method at t 3.2 is smaller than optimal Connectivity of k-means, hierarchal and SOM.

GDS2002: The best clusters are obtained at t = 11 where the Connectivity is minimum; i.e., 5.426191 and 5468 genes are obtained in two clusters. All Connectivity values of the proposed method are smaller than optimal Connectivity of k-means, hierarchal and SOM.

GDS2003: The best clusters are obtained at t = 9.5 where the Connectivity is minimum; i.e., 8.527381 and 4309 genes are obtained in two clusters. The Connectivity of the proposed method at t 9.5, 9.4 and 9.8 are smaller than optimal Connectivity of k-means, hierarchal and SOM.

GDS2196: The best clusters are obtained at t = 8.8 where the Connectivity is minimum; i.e., 10.70873 and 4637 genes are obtained in three clusters. The Connectivity of the proposed method at t 8.8 and 8.9 are smaller than optimal Connectivity of k-means, hierarchal and SOM.

GDS2712: The best clusters are obtained at t = 6.7 where the Connectivity is minimum; i.e., 70.65595 and 7824 genes are obtained in two clusters.

GDS2713: The best clusters are obtained at t = 5 where the Connectivity is minimum; i.e., 143.719 and 7824 genes are obtained in three clusters.

The above observations, based on Connectivity, indicate that the proposed method performs better on six data set in comparison with k-means, four data set in comparison with hierarchal and seven data set in comparison with SOM.

The observations from the Table 4 are as follows:

GDS608: The best clusters are obtained at t = 5 where the Silhouette is maximum; i.e., 0.443244 and 3998 genes are obtained in two clusters. The Silhouette of the proposed method at t 5, 5.5 and 5.6 are greater than optimal Silhouette of k-means, hierarchal and SOM.

GDS759: The best clusters are obtained at t = 6.9 where the Silhouette is maximum; i.e., 0.701075 and 2499 genes are obtained in two clusters. The Silhouette of the proposed method at t 6.9, 6.4, 6.7 and 6.5 are greater than optimal Silhouette of k-means, hierarchal and SOM.

GDS1013: The best clusters are obtained at t = 6.7 where the Silhouette is maximum; i.e., 0.928403 and 4635 genes are obtained in five clusters. The Silhouette of the proposed method at t 6.7 is greater than optimal Silhouette of k-means, hierarchal and SOM.

GDS1550: The best clusters are obtained at t = 6.2 where the Silhouette is maximum; i.e., 0.923376 and 4636 genes are obtained in three clusters. All ten Silhouette values of the proposed method are greater than optimal Silhouette of k-means, hierarchal and SOM.

GDS1611: The best clusters are obtained at t = 3.2 where the Silhouette is maximum; i.e., 0.573747 and 4637 genes are obtained in two clusters. The Silhouette of the proposed method at t 3.2 is greater than optimal Silhouette of k-means and SOM.

GDS2002: The best clusters are obtained at t = 11 where the Silhouette is maximum i.e. 0.889162 and 5468 genes are obtained in two clusters. All ten Silhouette values of the proposed method are greater than optimal Silhouette of k-means, hierarchal and SOM.

GDS2003: The best clusters are obtained at t = 0.3 where the Silhouette is maximum; i.e., 0.945976 and 78 genes are obtained in 32 clusters. However, genes in clusters are very small; and we consider some more clusters with moderate Silhouette value. The Silhouette of the proposed method at t 0.3, 9.5, 9.4 and 9.8 are greater than optimal Silhouette of k-means, hierarchal and SOM.

GDS2196: The best clusters are obtained at t = 8.2 where the Silhouette is maximum; i.e., 0.952399 and 4636 genes are obtained in three clusters. The Silhouette of the proposed method at t 8.2, 9, 8.3, 8.8, 8.9 and 8.5 are greater than optimal Silhouette of k-means and SOM.

GDS2712: The best clusters are obtained at t = 6.7 where the Silhouette is maximum; i.e., 0.845303 and 7824 genes are obtained in two clusters. The Silhouette of the proposed method at t 6.7, 6.9, 6.4, 6.8, 6.6 and 7 are greater than optimal Silhouette of k-means, hierarchal and SOM.

GDS2713: The best clusters are obtained at t = 4.9 where the Silhouette is maximum; i.e., 0.754239 and 7824 genes are obtained in three clusters. The Silhouette of the proposed method at t 4.9, 5, 4.4 and 4.5 are greater than optimal Silhouette of k-means, hierarchal and SOM.

The above observations, based on Silhouette, prove that the proposed method is better performed on all data set in comparison with k-means and SOM and better performed on eight data set in comparison with the hierarchal algorithm.

4 Conclusion

In this paper, a new method is proposed for clustering gene expression data that clusters genes and segregates functionally inactive genes or noise. The method assigns genes to a cluster if their expression level difference with a specific gene is less than a threshold t in each experimental condition. It works on column-wise distance calculation approach. The method is applied to ten yeast gene expression data. For evaluation and validation of the results, three clustering validation indices, Dunn, Connectivity and Silhouette, are used. The method is compared with k-means, hierarchal and SOM. The results of the proposed method applied to ten yeast cell cycle data show that it performed significantly better than k-means, hierarchal and SOM.

References

1. Tavazoie, S., Hughes, J.D., Campbell, M.J., Cho, R.J., Church, G.M.: Systematic determination of genetic network architecture. Nat. Genet. **22**(3), 281–285 (1999)
2. Dembele, D., Kastner, P.: Fuzzy c-means method for clustering microarray data. Bioinformatics **19**(8), 973–980 (2003)
3. Eisen, M.B., Spellman, P.T., Brown, P.O., Botstein, D.: Cluster analysis and display of genome-wide expression patterns. Proc. Nat. Acad. Sci. USA. **95**(25), 14863–14868 (1998)
4. Tamayo, P., Slonim, D., Mesirov, J., Zhu, Q., Kitareewan, S., Dmitrovsky, E., Lander, E.S., Golub, T.R.: Interpreting patterns of gene expression with self-organizing maps: Methods and application to hematopoietic differentiation. Proc. Nat. Acad. Sci. USA. **96**(6) 2907–2912 (1999)
5. Sharan, R., Shamir, R., CLICK: A clustering algorithm with applications to gene expression analysis. In: Proceedings of the Intelligent Systems for Molecular (ISMB), pp. 307–316 (2000)
6. Bandyopadhyay, S., Mukhopadhyay, A., Maulik, U.: An improved algorithm for clustering gene expression data. Bioinformatics **23**(21), 2859–2865 (2007)
7. Jiang, D., Tang, C., Zhang, A.: Cluster analysis for gene expression data: a survey. IEEE Trans. Knowl. Data Eng. **16**(11), 1370–1386 (2004)
8. Kerr, G., Ruskin, H.J., Crane, M., Doolan, P.: Techniques for clustering gene expression data. Comput. Biol. Med. **38**(3), 283–293 (2008)
9. Dunn, J.C.: Well-separated clusters and optimal fuzzy partitions. J. Cybern. **4**(1), 95–104 (1974)
10. Bolshakova, N., Azuaje, F.: Cluster validation techniques for genome expression data. Signal Process. **83**(4), 825–833 (2003)
11. Rousseeuw, P.J.: Silhouettes: a graphical aid to the interpretation and validation of cluster analysis. J. Comput. Appl. Math. **20**, 53–65 (1987)
12. Brock, G., Pihur, V., Datta, S., Datta, S.: clValid, an R package for cluster validation. J. Stat. Softw (Brock et al. March 2008) (2011)
13. Maji, P., Paul, S.: Rough-fuzzy clustering for grouping functionally similar genes from microarray data. IEEE/ACM Trans. Comput. Biol. Bioinf. **10**(2), 286–299 (2013)
14. Nieweglowski, L., Nieweglowski, M.L.: Package 'clv' (2015)

Detection of Moving Objects in Video Using Block-Based Approach

Amlan Raychaudhuri, Satyabrata Maity, Amlan Chakrabarti and Debotosh Bhattacharjee

Abstract In this paper, an efficient technique has been proposed to detect moving objects in the video under dynamic as well as static background condition. The proposed method consists block-based background modelling, current frame updating, block processing of updated current frame and elimination of background using bin histogram approach. Next, enhanced foreground objects are obtained in the post-processing stage using morphological operations. The proposed approach effectively minimizes the effect of dynamic background to extract the foreground information. We have applied our proposed technique on Change Detection CDW-2012 dataset and compared the results with the other state-of-the-art methods. The experimental results prove the efficiency of the proposed approach compared to the other state-of-the-art methods in terms of different evaluation metrics.

Keywords Moving object detection · Dynamic background · Background modelling · Block processing · Background elimination · Bin histogram

Biographical notes: This paper is a revised and expanded version of a paper entitled [**Moving Object Detection in Video under Dynamic Background Condition using Block-based Statistical Features**] presented at [**CICBA-2017, Calcutta Business School—West Bengal—India, March 24–25, 2017**].

A. Raychaudhuri (✉)
Department of Computer Science & Engineering, B. P. Poddar Institute
of Management & Technology, 137, VIP Road, Kolkata 700052, India
e-mail: amlanrc@gmail.com

S. Maity · A. Chakrabarti
A. K. Choudhury School of Information Technology, University of Calcutta, JD-2,
JD Block, Sector-III, Kolkata 700098, India
e-mail: satyabrata.maity@gmail.com

A. Chakrabarti
e-mail: amlanc@ieee.org

D. Bhattacharjee
Department of Computer Science & Engineering, Jadavpur University,
188, Raja S. C. Mallick Road, Kolkata 700032, India
e-mail: debotoshb@hotmail.com

© Springer Nature Singapore Pte Ltd. 2019
J. K. Mandal et al. (eds.), *Advances in Intelligent Computing*,
Studies in Computational Intelligence 687,
https://doi.org/10.1007/978-981-10-8974-9_8

1 Introduction

Moving object detection from a given video sequence under dynamic background condition is a very challenging task in video processing and has been an active research area in the field of computer vision for the last few decades [1, 2]. It has wide application in video surveillance, event detection, dynamic scene analysis, activity recognition and activity based human recognition [1, 2]. To detect moving objects from a given video sequence, the object regions which are moving are needed to be identified with respect to their background [2]. Moving object detection is done mainly using three different kinds of approaches [3]: background subtraction [4, 5], temporal differencing [6] and optical flow [7]. Two important steps for background subtraction approaches are: proper generation and updating of reference background model, and then application of an appropriate elimination technique to eliminate the background model from the current frame. Over the years, numerous different techniques have been proposed by various researchers. For all these methods, the most important steps are generation of background model and how it is updated over time.

Detection of moving objects remains an open research problem even after research of several years in this field. To obtain an accurate, robust and high-performance approach is still a very challenging job. Shaikh et al. [3] mentioned some of the challenges of background subtraction related to video surveillance in their work. The challenges are—(a) Illumination changes: for a video sequence, illumination may change gradually or in some cases rapidly. To detect a moving object accurately from these kinds of videos, the background model should take this into consideration. (b) Dynamic background: in some videos, the background is not static. It contains movement (a fountain, wave of water, movement of tree leaves, etc.) in some regions of the background. To eliminate these kinds of movement is a challenging task. (c) Camouflage: some portion of the objects' intensity is very much similar to that of the background intensity in the same region. To classify objects and background correctly under this scenario is a huge challenging task. (d) Presence of shadows: shadows are created mainly due to presence of foreground objects; removal of shadows occurring due to foreground objects is a very challenging task.

Background subtraction is a popularly used method for motion segmentation in static scenes [8]. Heikkila and Silven [9] used the simple version of this scheme. They have tried to detect moving regions by pixel-wise subtraction of the current frame from a background reference frame. The pixels will be treated as foreground if the corresponding difference is above a threshold. The generation of the background image by applying some technique is known as background modelling. After creating a foreground pixel mask, some morphological operations are used such as erosion, dilation and closing for noise reduction and to enhance the detected foreground regions in the post-processing step. The reference background is updated over time to handle the dynamic scene changes.

Some authors used temporal differencing to detect moving regions by taking the pixel-by-pixel difference between consecutive two or three frames in a video sequence [10–13]. Lipton et al. [10] proposed a two-frame differencing method where the pixel-by-pixel difference is calculated for two consecutive frames. If that difference is above some threshold value, then the corresponding pixel is marked as foreground. In [11], the authors have used three frames differencing in order to overcome the shortcomings of two frames differencing. For instance, Collins et al. [12] developed a hybrid method which combines three frames differencing with an adaptive model for background subtraction. The hybrid algorithm successfully discriminates moving regions from the background region in the video without the limitations of temporal differencing and background subtraction.

In [3], the authors proposed a low-cost background subtraction method for detecting moving objects under dynamic background conditions. They have created a background model by computing the median of corresponding pixels of a certain number of frames. Then through block processing and using statistical method, background is eliminated from the current frame. After that in the post processing step, some morphological operations are applied to obtain the foreground object.

Kumar and Yadav [14] proposed a method to realize the foreground moving blobs with the help of suitable initialization and updating of the background model for object tracking. An initial motion field is generated by applying spatial-temporal filtering on the consecutive frames. Then, for the pixels above a certain value of the difference image, block-wise entropy is calculated to get the moving object.

Rahman et al. [15] proposed a method for improving traditional background subtraction (BGS) techniques by using a gradient-based edge detector, named second derivative in gradient direction (SDGD) filter with the BGS output.

Stauffer and Grimson [16] proposed a statistical method based on an adaptive background mixture model for real-time tracking. There, every pixel is separately modelled by a mixture of Gaussians which are updated using incoming image data. For detection of a pixel belongs to a foreground or background, the Gaussian distributions of the mixture model for that pixel are evaluated. Heras et al. [17] have used Gaussian mixture models for background subtraction. They have defined a splitting operation based on the results of Split and Merge EM algorithm for background subtraction. Varadarajan et al. [18] worked on dynamic background environment. Under this environment, modelling pixels using Gaussian mixture model is not very effective. So, they model region of pixels as mixture distributions rather than a single pixel.

Tiefenbacher et al. [19] proposed the proportional-integral-derivative (PID) controllers which can regulate the decision threshold of the background dynamics as well as update rate of the background model. For proper foreground and background segmentation, handling the dynamics of background is very essential.

Morde et al. [20] proposed a system having background model based on Chebyshev probability inequality. The system improves the segmentation accuracy using shadow detection and significance feedback from higher-level object classification.

In [7, 21, 22], the authors have used optical flow methods to make use of the flow vectors of the objects which are moving over time for detection of moving objects in an image. In this approach, they have computed the apparent velocity and the direction of every pixel. It is an effective but time-consuming method.

In our research work, we have detected moving objects in a video under dynamic as well as static background condition. After generating a background model using block processing, efficiently the background is eliminated from the current frame using block processing and thresholding of bin histogram. Finally at the post-processing step, we applied some morphological operations to eliminate very small wrongly detected foreground regions and enhance the original foreground or moving regions. The remainder of this paper is presented as follows: Sect. 2 yields the details of proposed method. The experimental results and analysis are discussed in Sect. 3. Finally, the conclusion is given in Sect. 4.

2 Proposed Method

In this research work, we have proposed a new technique to detect moving objects in a video under dynamic as well as static background condition. In our previous work [23], we have detected moving objects only for videos under dynamic background condition.

Here, we have worked on gray-scale frames. Thus, at first, all the frames of a video sequence are extracted, and they are converted into gray-scale images. The brief sketch of our proposed method is shown in Fig. 1. The proposed method has the following important steps: (a) background modelling with block processing, (b) averaging the background frames to create a background image, (c) updating current frame, (d) block processing of the current frame and background elimination and (e) post-processing operations.

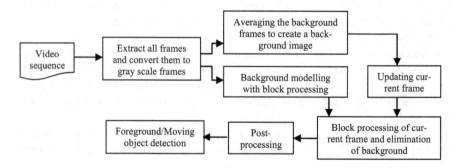

Fig. 1 A brief sketch of the proposed method

2.1 Background Modelling with Block Processing

To detect the moving objects in the current frame, we have generated a background model consisting of previous n frames where the objects are not present. The value of n should be large enough as to generate a realistic background model for the dynamic background. We have taken recent frames for consideration of background model to cater the recent changes in the scene.

At first, each frame (*frame size* $= r \times c$) of these n frames is divided into a number of equal sized logical blocks bg_{ij} (*block size* $= m \times m$), resulting $i \times j$ number of blocks, where $i = 1$ *to* r/m and $j = 1$ *to* c/m. Block-wise we gather the statistical features of the background. For block processing, bin histogram is used here for each of the blocks. The pixels, whose intensity values are close to each other, belong to the same bin depending on the bin size. By using bin histogram, we efficiently handle the dynamic nature of the background. There are total 256 intensity values (0–255) in a gray-scale image. Depending on the bin size (*bs*), a total number of bins (*bn*) are calculated using Eq. (1).

$$bn = 256/bs \qquad (1)$$

Thus, total bn number of bins are considered for each block and corresponding to those bins, a bin histogram is computed which is shown in Fig. 2. A block bg_{ij} of a background frame is shown in Fig. 2a. Its corresponding histogram and bin histogram are shown in Fig. 2b and Fig. 2c respectively. In Fig. 2c, $bs = 16$ is considered, so as a result the value of bn is also 16. A bin histogram represents the

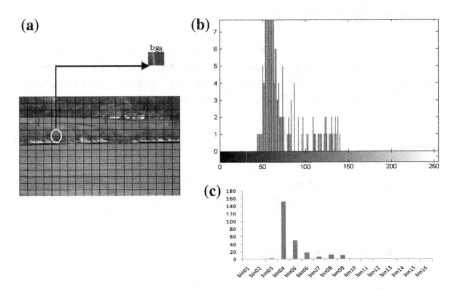

Fig. 2 **a** Block bg_{ij} of a background frame, **b** Histogram of block bg_{ij}, **c** Bin histogram of block bg_{ij}

frequency of bins in a block. If the intensity of a pixel x_t is represented by *intensity* (x_t), then the bin at which pixel x_t will be mapped is given by Eq. (2). This computation is done for all the blocks of these n frames. Thereafter, a background model is generated by averaging the bin histogram values corresponding to each block of these n frames, which is shown in Eq. (3).

$$x_t \in bn_l, \quad if \ bn_l = int(intensity(x_t)/bs) + 1$$
$$\{\forall t = 1 \ to \ m \times m, \ 1 \le l \le bn\} \tag{2}$$

where x_t pixel maps to bin bn_l, bs is the bin size and bn represents the total number of bins in the bin histogram, total $(m \times m)$ number of pixels are present in each block.

$$bin\_hist(bg_{ij})_{bg} = average \ \{bin\_hist(bg_{ij})_1, \ bin\_hist(bg_{ij})_2, \ \ldots, \ bin\_hist(bg_{ij})_n\}$$
$$\{\forall i = 1 \ to \ r/m, \ \forall j = 1 \ to \ c/m\}$$

$$\tag{3}$$

where $bin\_hist(bg_{ij})_k$ means bin histogram of block bg_{ij} (size $= m \times m$) of k-th frame and $bin\_hist(bg_{ij})_{bg}$ represents bin histogram of block bg_{ij} of the background model, n represents the total number of frames considered for generating background model, r and c denote the number of rows and the number of columns of a frame.

2.2 Averaging the Background Frames to Create a Background Image

In this step, a background image (BG) is created with the help of same n number of background frames which are used in the previous step. The pixel intensity of each position of BG is calculated by averaging the intensity values of the corresponding position for the n number of frames, which is shown in Eq. (4).

$$BG(i,j) = \sum_{k=1}^{n} B_k(i,j)/n \qquad \{\forall i = 1 \ to \ r, \ \forall j = 1 \ to \ c\} \tag{4}$$

here $BG(i, j)$ and $B_k(i, j)$ represent the (i, j)-th pixel intensity of the background image (BG) and the k-th background frame respectively.

2.3 Updating Current Frame

Before block processing of the current frame, some pre-processing is done on the current frame. Here we have worked on the video sequences under dynamic and static background condition. For dynamic background, it is changing over time due to various causes (a fountain, wave of water, movement of tree leaves, etc.). Under this situation, it is very difficult to extract accurate moving objects from the video sequence. For static background, it may be happened that, the intensity values are changed due to illumination changes. So, to handle these kinds of situations, the current frame (CR) is modified with the Eq. (5).

$$
\begin{aligned}
CR(i,j) &= BG(i,j), \quad if \, |\, CR(i,j) - BG(i,j)\,| < th_1 \\
&= CR(i,j), \quad otherwise \quad \{\forall i = 1 \; to \; r, \forall j = 1 \; to \; c\}
\end{aligned}
\tag{5}
$$

For the current frame (CR), the intensity value of a pixel is modified by the corresponding background pixel intensity, if the absolute difference between these two pixels intensity is less than a threshold value (th_1). Otherwise, no change for the pixel value of the current frame. Here computation of th_1 value is shown in Eq. (6).

$$
th_1 = std(CR)/2 - \gamma
\tag{6}
$$

where std(CR) means the standard deviation of the current frame (CR) and γ is parameter which is taken the value 5 here.

By this way we have modified the current frame (CR) and then in the next step we have used the modified current frame (MCR).

2.4 Block Processing of the Current Frame and Background Elimination

In this step, the modified current frame (MCR) is also divided into same number of equal-sized logical blocks mcr_{ij} (block size = $m \times m$), resulting $i \times j$ number of blocks, where $i = 1$ to r/m and $j = 1$ to c/m. Then, each block of the frame MCR (frame size = $r \times c$) is compared with the corresponding block of the background model on the basis of bin histogram. The formula for classify a pixel x_t into object or background is given in Eq. (7).

$$
\begin{aligned}
I_1(mcr_{ij}(x_t)) &= 1, \quad if \, freq(bn_l) < th_2 \quad \{\forall x_t \in bn_l, \; 1 \leq l \leq bn\} \\
&= 0, otherwise \quad\quad \{\forall i = 1 \; to \; r/m, \forall j = 1 \; to \; c/m\}
\end{aligned}
\tag{7}
$$

A pixel x_t of a particular block mcr_{ij} of the modified current frame will be treated as foreground pixel, if its corresponding bin's frequency $freq(bn_l)$ is less than a threshold value th_2 in the bin histogram of the corresponding block bg_{ij} of the background model. Otherwise, the pixel x_t will be treated as background. In this way, after processing all the pixels of each block of the modified current frame, a new binary image I_l is created, where all the foreground or moving object pixels are assigned with value '1' and background pixels are with '0'.

We have compared the bin histogram of a block of the current frame with the bin histogram of the corresponding block of the background model using a threshold th_2 for all the blocks. The threshold value th_2 is dependent on the video sequences. In this way, we have efficiently eliminated the dynamic nature of the background information from the current frame.

2.5 Post-processing Operations

After background elimination by the above mentioned method, a binary image I_l is generated. This image contains some noises outside the foreground region and some foreground pixels are misclassified as background pixels. So in the post-processing step, we have applied two morphological operations—*open* and *close* to reduce noises and misclassifications.

The morphological *open* operation is applied on the binary image I_l to reduce noises outside the foreground region. In this case, very small regions which are misclassified as foreground regions are eliminated from the image I_l. Then in the resultant image, some portion of the foreground region may be misclassified as background. Thus to obtain enhanced foreground object regions, the morphological closing operation is applied on the resultant image.

In this way, the desired binary image is obtained. Where white pixels represent foreground or moving object and black pixels represent background. Figure 3 shows the results for the sample frames.

Fig. 3 Results on sample frames. **a** *boats* frame in RGB color, its gray image, background frame and output **b** *highway* frame in RGB color, its gray image, background frame and output

3 Experimental Results and Analysis

To evaluate the effectiveness of our approach, we performed our experiments on the Change Detection *CDW-2012 (dynamic background and baseline)* [24] dataset. The *dynamic background* dataset is a standard benchmark dataset consisting of six video categories with a dynamic background. The *baseline* dataset is also a standard benchmark dataset consisting of four video categories. The threshold value th_2 which is described in Sect. 2.4, is chosen based on experimental testing. Here for these datasets we have considered the parameter th_2 value in the range from 5 to 10 to obtain the best results. Tables 1 and 2 show the parameter value th_2 chosen for the different video sequences of *Dynamic background* dataset and *Baseline* dataset respectively. For effectiveness of our research work, we have compared our results with the readily available ground truth data in the above mentioned datasets.

To evaluate the performance of our proposed method, different evaluation measures techniques [24] have been applied here. The evaluation measures are based on the following parameters: TP = true positives; the number of foreground pixels correctly classified as foreground. TN = true negatives; the number of background pixels correctly classified as background. FP = false positives; the number of background pixels misclassified as foreground. FN = false negative; the number of foreground pixels misclassified as background. The different metrics are described below.

Recall: It is defined as the ratio of true positive to true positive and false negative. A high value of recall is desired.

$$Recall = TP/(TP + FN) \tag{8}$$

Specificity: It is defined as the ratio of true negative to true negative and false positive. A high value of this metric is desired.

Table 1 The parameter th_2 chosen for different video sequences of *Dynamic background* dataset

Videos	th_2
Boats	5
Canoe	10
Fountain01	6
Fountain02	10
Fall	6
Overpass	8

Table 2 The parameter th_2 chosen for different video sequences of *Baseline* dataset

Videos	th_2
Pedestrians	10
Highway	6
PETS2006	8
Office	5

$$Specificity = TN/(TN + FP) \tag{9}$$

False Positive Rate (FPR): It is the ratio of false positive to false positive and true negative. A low value of FPR is desired.

$$FPR = FP/(FP + TN) \tag{10}$$

False Negative Rate (FNR): It is the ratio of false negative to false negative and true positive. A low value of FNR is desired.

$$FNR = FN/(FN + TP) \tag{11}$$

Percentage of the Wrong Classification (PWC): It is 100 times the ratio of false positive and false negative to all the detected pixels. It is better if the PWC value is lower.

$$PWC = 100 * (FP + FN)/(TP + TN + FP + FN) \tag{12}$$

F-measure: It is based on recall and precision both. A high value of F-measure is desired.

$$F - measure = 2 * Recall * Precision/(Recall + Precision) \tag{13}$$

Precision: It is the ratio of true positive to true positive and false positive. A high value of Precision is desired.

$$Precision = TP/(TP + FP) \tag{14}$$

3.1 Dynamic Background

The *dynamic background* dataset [24] is a standard benchmark dataset consisting of six video categories with a dynamic background. Figure 4 shows sample frames for each category of videos of *dynamic background* dataset, their corresponding ground truth and outputs by applying our proposed method. Table 3 shows the average results of the proposed method for all the six video categories based on the above mentioned seven metrics.

In Table 4, the comparison results of our method with the other state-of-the-art methods using an average of all seven evaluation metrics are shown. It can be seen that, in most of the cases, our proposed method gives the best or the 2nd best result

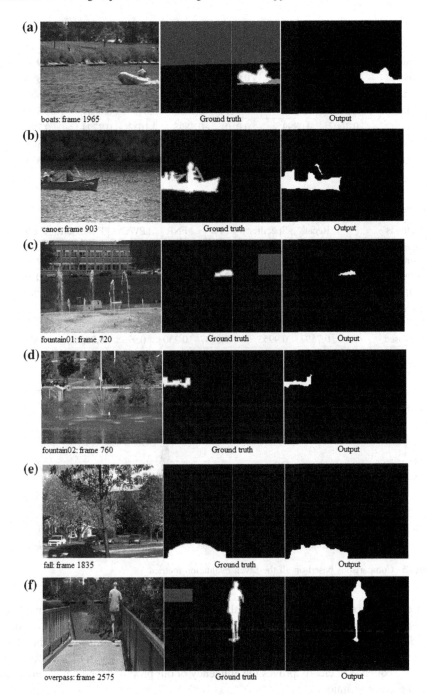

Fig. 4 Sample frames of *dynamic background* dataset, corresponding ground truths and outputs of the proposed method for six video categories (**a–f**)

Table 3 Average results on all the videos of *dynamic background*

Videos	Recall	Specificity	FPR	FNR	PWC	F-measure	Precision
Boats	0.9333	0.9979	0.0021	0.0667	0.4798	0.9419	0.9507
Canoe	0.9363	0.9923	0.0077	0.0637	1.1530	0.9173	0.8990
Fountain01	0.5730	0.9998	0.0002	0.4270	0.3650	0.5888	0.6055
Fountain02	0.5647	0.9990	0.0010	0.4353	0.6406	0.6881	0.8804
Fall	0.7602	0.9899	0.0101	0.2398	2.1632	0.7794	0.7996
Overpass	0.8506	0.9931	0.0069	0.1494	1.4059	0.8583	0.8661
Average	0.7697	0.9953	0.0047	0.2303	0.9798	0.7956	0.8336

Table 4 Comparison results with the other state-of-the-art methods

Methods	Recall	Specificity	FPR	FNR	PWC	F-measure	Precision
Stauffer and Grimson [16]	0.7108	0.9860	0.0140	0.2892	3.1037	0.6624	0.7012
PBAS [25]	**0.7840**	0.9898	0.0102	**0.2160**	1.7693	0.7532	0.8160
CDPS [26]	0.7769	0.9848	0.0152	0.2231	2.2747	0.7281	0.7610
Tiefenbacher et al. [19]	0.7542	**0.9989**	**0.0011**	0.2458	**0.4624**	0.7357	0.8291
Proposed	0.7697	0.9953	0.0047	0.2303	0.9798	**0.7956**	**0.8336**

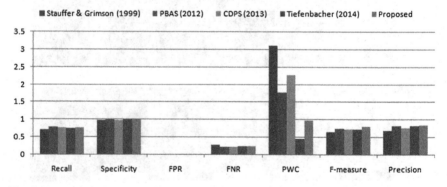

Fig. 5 Comparisons based on all the seven evaluation metrics

based on the seven metrics. Thus it validates the efficiency of our proposed method. Figure 5 presents graphically the comparison results based on all the seven metrics.

Figure 6 illustrates the graphical analysis of our method compared to the other state-of-the-art methods [16, 19, 25, 26] in respect to all the seven metrics for each video sequence. It clearly proves the efficiency of our proposed technique compared to the other techniques.

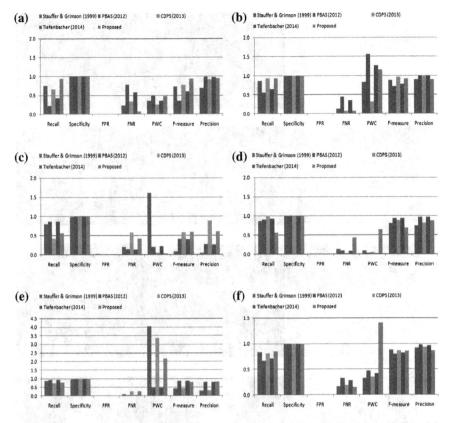

Fig. 6 Comparison analysis for six videos: **a** boats, **b** canoe, **c** fountain01, **d** fountain02, **e** fall, and **f** overpass

3.2 Baseline

We have also applied our proposed method to the *Baseline* dataset [24] video sequence. It is a standard benchmark dataset consisting of four video categories. In those video sequences, though background is static, but it has been changed due to illumination change. In some sequences, the movement of foreground object is very slow. The results obtained here for the dataset are very impressive. Figure 7 shows sample frames for each category of videos of baseline dataset, their corresponding ground truth and outputs by applying our proposed method. Table 5 shows the average results of the proposed method for all the four video categories based on the above mentioned seven metrics.

In Table 6, the comparison results of our method with the other state-of-the-art methods using an average of all seven evaluation metrics are shown. It can be seen that, in most of the cases, our proposed method provides the best result based on the

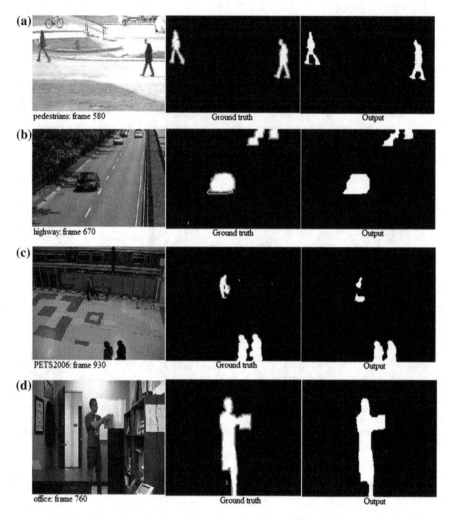

Fig. 7 Sample frames of *baseline* dataset, corresponding ground truths and outputs of the proposed method for four video categories (**a–d**)

Table 5 Average results on all the videos of *Baseline*

Videos	Recall	Specificity	FPR	FNR	PWC	F-measure	Precision
Pedestrians	0.9138	0.9985	0.0015	0.0862	0.3208	0.9199	0.9262
Highway	0.9313	0.9945	0.0055	0.0687	1.0001	0.9303	0.9293
PETS2006	0.7869	0.9971	0.0029	0.2131	0.6117	0.7965	0.8064
Office	0.9021	0.9967	0.0033	0.0979	1.0850	0.9300	0.9597
Average	0.8836	0.9967	0.0033	0.1164	0.7544	0.8942	0.9054

Table 6 Comparison results with the other state-of-the-art methods

Methods	Recall	Specificity	FPR	FNR	PWC	F-measure	Precision
Morde et al. [20]	0.8266	0.997	0.003	0.1734	0.8304	0.8646	**0.9143**
SGMM [17]	0.868	0.9949	0.0051	0.132	1.2436	0.8594	0.8584
Stauffer and Grimson [16]	0.818	0.9948	0.0052	0.182	1.5325	0.8245	0.8461
Varadarajan et al. [18]	0.7082	**0.9981**	**0.0019**	0.2918	1.5935	0.7848	0.9125
Proposed	**0.8836**	0.9967	0.0033	**0.1164**	**0.7544**	**0.8942**	0.9054

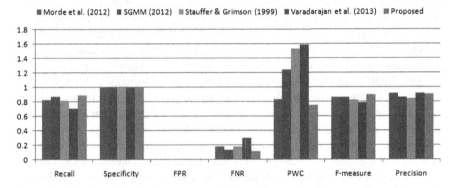

Fig. 8 Comparisons based on all the seven evaluation metrics

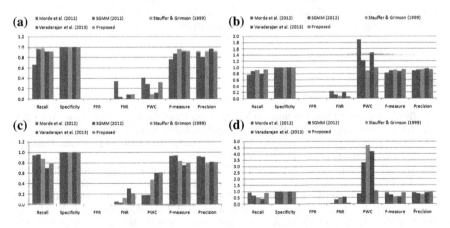

Fig. 9 Comparison analysis for four videos: **a** pedestrians, **b** highway, **c** PETS2006, and **d** office

seven metrics. Thus it validates the efficiency of our proposed method. Figure 8 presents graphically the comparison results based on all the seven metrics.

Figure 9 describes the graphical analysis of our method compared to the other state-of-the-art methods [16–18, 20] in respect to all the seven metrics for each video sequence. It clearly proves the efficiency of our proposed technique compared to the other techniques.

4 Conclusion

In this paper, we propose an efficient technique to detect moving objects in a video which works efficiently for the two contrasting categories of video i.e. either having dynamic background or having almost static background. In our work, we have developed an efficient block-based background modelling technique along with block based processing of the modified version of the current frame, where the current frame is updated based on the background information. Background is eliminated by comparing bin histogram of the corresponding blocks using thresholding a strategy. We have executed our work on a standard benchmark dataset, and the results are excellent for majority of the video sequences. Our result is very competitive with the other state-of-the-art methods. In most of the cases, we overcome the challenges like—illumination change and dynamic background satisfactorily. However, in this method, we could not handle shadow detection and camouflage very well. In future, we will work to resolve these two challenges.

Acknowledgements The authors of this paper would like to acknowledge the website (www. changedetection.net) for obtaining the Change Detection *CDW-2012 (dynamic background and baseline)* dataset to perform their research work effectively.

References

1. Bovik, A.C.: Hand Book on Image and Video Processing. Academic Press, New York (2000)
2. Tekalp, A.M.: Digital Video Processing. Prentice Hall, New Jersey (1995)
3. Shaikh, S.H., Saeed, K., Chaki, N.: Moving object detection using background subtraction. In: Springer Briefs in Computer Science (2014)
4. Piccardi, M.: Background subtraction techniques: a review. In: Proceedings IEEE International Conference on Systems, Man and Cybernetics (SMC), vol. 4, pp. 3099–3104 (2004)
5. Elhabian, S., El-Sayed, K., Ahmed, S.: Moving object detection in spatial domain using background removal techniques — state-of-art. Recent Patents Comput. Sci. **1**, 32–54 (2008)
6. Fejes, S., Davis, L.S.: Detection of independent motion using directional motion estimation. Comput. Vis. Image Understand. (CVIU) **74**(2), 101–120 (1999)
7. Paragios, N., Deriche, R.: Geodesic active contours and level sets for the detection and tracking of moving objects. IEEE Trans. Pattern Anal. Mach. Intell. **22**(3), 266–280 (2000)
8. McIvor, A.M.: Background subtraction techniques. In: Proceedings of Image and Vision Computing (2000)

9. Heikkila, J., Silven, O.: A real-time system for monitoring of cyclists and pedestrians. Proc. Image Vis. Comput. **22**(7), 563–570 (2004)

10. Lipton, A.J., Fujiyoshi, H., Patil, R.S.: Moving target classification and tracking from real time video. In: Proceedings 4th IEEE Workshop on Applications of Computer Vision (WACV '98), pp. 129–136 (1998)

11. Wang, L., Hu, W., Tan, T.: Recent developments in human motion analysis. Pattern Recogn. **36**(3), 585–601 (2003)

12. Collins, R.T., Lipton, A.J., Kanade, T., Fujiyoshi, H., Duggins, D., Tsin, Y., Tolliver, D., Enomoto, N., Hasegawa, O., Burt, P., Wixson, L.: A system for video surveillance and monitoring. In: Technical Report CMU-RI-TR-00-12. The Robotics Institute, Carnegie Mellon University (2000)

13. Haritaoglu, I., Harwood, D., Davis, L.S.: W^4: Real-time surveillance of people and their activities. IEEE Trans. Pattern Anal. Mach. Intell. **22**(8), 809–830 (2000)

14. Kumar, S., Yadav, J.S.: Video object extraction and its tracking using background subtraction in complex environments. Perspect. Sci. **8**, 317–322 (2016)

15. Rahman, A.Y.F., Hussain, A., Zaki, W., Zaman, B., Tahir, M.: Enhancement of background subtraction techniques using a second derivative in gradient direction filter. J. Electr. Comput. Eng. **2013**, 1–12 (2013)

16. Stauffer, C., Grimson, W.E.L.: Adaptive background mixture models for real-time tracking. In: Proceedings IEEE Computer Society Conference on Computer Vision and Pattern Recognition (CVPR), vol. 2 (1999)

17. Heras Evangelio, R., Pätzold, M., Sikora, T.: Splitting Gaussians in mixture models. In: Proceedings of the 9th IEEE International Conference on Advanced Video and Signal-Based Surveillance (2012)

18. Varadarajan, S., Miller, P., Huiyu Z.: Spatial mixture of Gaussians for dynamic background modeling. In: Proceedings of the 10th IEEE International Conference on Advanced Video and Signal Based Surveillance (AVSS '13), pp. 63–68 (2013)

19. Tiefenbacher, P., Hofmann, M., Merget, D., Rigoll, G.: PID-based regulation of background dynamics for foreground segmentation. In: Proceedings of Image Processing (ICIP), pp. 3282–3286 (2014)

20. Morde, A., Ma, X., Guler, S.: Learning a background model for change detection. In: Proceedings of IEEE Workshop on Change Detection (2012)

21. Wixson, L.: Detecting salient motion by accumulating directionally—consistent flow. IEEE Trans. Pattern Anal. Mach. Intell. **22**(8) (2000)

22. Pless, R., Brodsky, T., Aloimonos, Y.: Detecting independent motion: the statistics of temporal continuity. IEEE Trans. Pattern Anal. Mach. Intell. **22**(8), 768–773 (2000)

23. Raychaudhuri, A., Maity, S., Chakrabarti, A., Bhattacharjee, D.: Moving object detection in video under dynamic background condition using block-based statistical features. In: Proceedings CICBA 2017, Part II, CCIS 776, pp. 371–383 (2017)

24. Goyette, N., Jodoin, P.-M., Porikli, F., Konrad, J., Ishwar, P.: Change detection.net: a new change detection benchmark dataset. In: Proceedings IEEE Workshop on Change Detection (CDW-2012) at CVPR (2012)

25. Hofmann, M., Tiefenbacher, P., Rigoll, G.: Background segmentation with feedback: the pixel-based adaptive segmenter. In: Proceedings IEEE Workshop on Change Detection, pp. 38–43 (2012)

26. Francisco, J., Lopez, H., Rivera, M.: Change detection by probabilistic segmentation from monocular view. Proc. Mach. Vis. Appl. **25**(5), 1175–1195 (2014)

Author Index

© Springer Nature Singapore Pte Ltd. 2019
J. K. Mandal et al. (eds.), *Advances in Intelligent Computing*,
Studies in Computational Intelligence 687,
https://doi.org/10.1007/978-981-10-8974-9

Printed in the United States
By Bookmasters